Progress and Values
in the Humanities

Progress and Values
in the Humanities:
Comparing Culture and Science

Volney Gay

COLUMBIA UNIVERSITY PRESS

NEW YORK

Columbia University Press
Publishers Since 1893
New York   Chichester, West Sussex
Copyright © 2010 Columbia University Press

Library of Congress Cataloging-in-Publication Data
Gay, Volney Patrick.
  Progress and values in the humanities : comparing culture and science /
Volney Gay.
    p.   cm.
  Includes bibliographical references and index.
  ISBN 978-0-231-14790-3 (cloth : alk. paper)—ISBN 978-0-231-51981-6 (e-book)
  1. Science and the humanities.   I. Title.
  AZ341.G39   2009
  001.3—dc22
                                                                2009024174

References to Internet Web sites (URLs) were accurate at the time of writing. Neither
the author nor Columbia University Press is responsible for URLs that may have
expired or changed since the manuscript was prepared.

For Gordon Gee

President, The Ohio State University

Mentor, Colleague, Friend

CONTENTS

ACKNOWLEDGMENTS

I am grateful to the Metanexus Foundation and the two executive directors with whom I have worked, William Grassie and Eric Weislogel. They sponsored a four-year research group on science and religion at Vanderbilt University, without which this book would not have appeared. Mark Edwards (Harvard Divinity School), Christof Koch (California Institute of Technology), and Ronald Grimes (Wilfrid Laurier University) read the manuscript and helped clarify my thinking. I am also grateful to Chancellor Gordon Gee, Provost Nicholas Zeppos, and Dean Richard McCarty of the College of Arts and Science at Vanderbilt for their support.

I am grateful to the following for permission to reprint images and photographs:

Figure I.1, Image of microscope, Allan Wissner, Antique-Microscope.com

Figure 4.2, cover of *DSM-IV-TR*, reprinted with permission from the *Diagnostic and Statistical Manual of Mental Disorders*, Text Revision, 4th edition (copyright 2000), American Psychiatric Association

Figure 6.1, Freud's diagram of auditory nerves, Sigmund Freud Copyrights/ Paterson Marsh Ltd.

Figure 6.2, Freud's illustration of ego and id, Sigmund Freud Copyrights/ Paterson Marsh Ltd.

*Progress and Values
in the Humanities*

INTRODUCTION

The campus is quiet in the morning. A university police car glides across an empty parking lot, the end of the night shift. Steam rises from my carryout coffee as I wait for the walk signal on 21st Avenue. I peer across the street at three buildings in front of me. On the right, against the gray sky, I see the spire of Vanderbilt Divinity School. It is a brick three-story building constructed in 1960. It houses some 30 professors and serves 180 divinity students and about 100 graduate students. Directly in front of me is Vanderbilt's central library—a brick building constructed in 1941. It is three stories tall, flanked by five-story towers; stuck onto it is an eight-story addition built in 1969. The library building serves the humanities and social sciences. Medicine and the sciences have separate library buildings.

To the left of the library is a huge, new structure—Medical Research Building III (MRB III). Nine stories tall and encompassing 350,000 square feet (more than the Divinity School and the central library combined), MRB III cost $95 million to build and opened in 2002; it joined the preexisting MRB I and MRB II. Before it opened, plans were already laid to construct MRB IV about a quarter mile away. MRB IV will cost $110 million and add another 200,000 square feet for biomedical research.

The last four divinity deans have tried to expand the divinity building and relieve overcrowding of faculty and students. Unhappily, no donor has yet answered their prayers. University librarians have also struggled to replace the main building. With much hard work, the current librarian convinced university administrators to consider refurbishing the main building or construct a new building. After hearing an architect present an exciting design, hopes ran high but then subsided—not enough money.

Why is this pattern, seen at every university, so common? Why do money, new buildings, and resources seem to flow like a river to the sciences and medicine yet barely trickle to divinity studies and other humanities? Are scientists, some of whom wear dark pants and white tennis shoes, more worthy than I and my fellow humanists? Are they smarter or more skilled than we? Some might be; but not all. Scientists and medical researchers do not attract vast riches because of their personalities or their IQs, as wonderful as those may be.

I cradle my coffee and let my envy subside. The walk signal flashes and I amble past the library. Through a lighted window I see a shelf of books in the Bible section and remember James, a former colleague, a specialist in the Hebrew Bible. In a long career, James received a dozen honorary doctorates, dominated his branch of study, taught generations of students, and graced our school. I cannot imagine a better person. Assuming that the ratio of saints to scoundrels is the same in the sciences as in the humanities, MRB III was not built to honor persons of unusual character. If any individual deserves reward, it is James, and the new wing of the Divinity School—if it's ever built—should be named for him.

This might happen. In the meantime, no doubt plans are underway for an MRB V. This book is about that fact. Medical research buildings are constructed because the U.S. government and private donors believe that pouring billions of dollars into medical research is a good thing. They (and I) believe that science will cure many diseases and alleviate suffering. This conviction is based on the fact that Western science has progressed in ways not matched by any other human endeavor. When medicine oriented itself toward the sciences in the nineteenth century, it joined this progressive movement. Medical science has not yet cured all forms of cancer. There is, however, no reason why it cannot do so in the future. This promise justifies the billions of dollars spent every year on that disease alone.

Progress in the sciences is normal and assumed to be continuous; progress in the humanities is rare and uncertain. My colleague James certainly made progress. He advanced his field of study; he helped us better understand Hebrew scripture. Yet, even James's success is partial; he did not convince everyone, and different schools of biblical studies dispute his work. Beyond this common feature of humanistic disciplines is the issue of the James's subject matter, the objects of his study.

I've reached my office. It's on the third floor of a pleasant, frumpy building on the oldest part of Vanderbilt's campus. I reheat my coffee—twenty seconds in the microwave in the utility room—and prepare my day. I will spend part of it teaching, part supervising psychiatry residents, part seeing psychoanalytic patients, and part dealing with the Center for the Study of Religion and Culture. The question of progress comes back to me. I flop my feet onto an old magazine on my desk and think about James and his subject matter.

HUMANISTS AND THEIR SUBJECT MATTERS

The objects that James studied are the thoughts of people who died about three thousand years ago. These objects are obscure in the ways that cancer is not. Cancer occurs because something happens inside normal cells to make them grow abnormally. One part of the cell sends faulty signals to another part. Tumors develop and death comes to the animal or person in whose body this has occurred. In principle, we can find ways to read these faulty signals, correct them, and stop the disease. Finding those signals and understanding how cells process them requires thousands of persons working for many years.

The objects of cancer research are complex, but not infinitely complex. Our scientific faith tells us that cancer cells cannot violate the laws of chemical interaction. This faith gives us a sorting device. With it we can cast out useless ideas and theories. It offers us a rule: if my theory of cancer includes the premise that cancer cells manufacture their own energy spontaneously (contrary to chemical laws), then my theory is no good. I will get no funding and no rooms in MRB III or in any other MRB on any other campus.

The objects of James's research, the things that he studied, are texts by Jewish authors, the ways in which later writers have reflected upon these earlier authors, and the cultural worlds in which these texts arose. These accounts and stories are complex; they are not infinitely complex—it seems. Hebrew scripture is of a limited size. Numerous brilliant persons have read, analyzed, compared, and studied each syllable of each text. Surely we can point to unqualified success in all this scholarship? If so, would that be progress in the humanities?

Here we feel the ground shift. Stories from the Hebrew Bible are about human beings and God. If human beings and God are like cancer cells, then over time we should be able to understand them fully. We could mount a scientific investigation of the Bible, of God and God's nature. We could look confidently toward volumes labeled *Progress in Religion* as we look annually for *Progress in Cell Biology* or *Progress in Neuroscience*.

We know that cancer is a natural event. Whatever causes it will yield to scientific investigation—eventually. What do we know about God and God's relationship to human beings? We know nothing except what we read in books like the Hebrew Bible, or in other scriptures of other faiths. To understand what Jewish theologians mean by "*God and human being*," we have to understand the terms *God* and *human being*. To understand those terms we have to rely on yet another story, another narrative about both partners in this duet. We are caught in a circle and we cannot escape.

But if we reject the Hebrew Bible and similar texts, we reject their subject matter—the relationship between human beings and God. There shall be no authority on human beings except that which emerges from the halls of science. We do not turn to priests for theories of cancer, why should we turn to them for theories of other human disorders? Here, though, we should stop and ponder what this decision costs us. It means that we reject the idea that there could be a spirit or intelligence superior to our own. This, in turn, means that we assign to humans determination of what is good and bad, right and wrong. Which human beings should decide this? Perhaps the National Academy of Sciences?

The National Academy of Sciences is an organization of two thousand scientists considered by their peers to be the best in their fields. They are the All Star Team of American scientists. If we asked this august group to decide questions about human beings and God, they would refuse. As honest scientists, they would inform us—brusquely—that these questions are not subject to scientific dispute because we cannot define either term. The term *God* is not definable— except, again, through religious traditions. The term *human being* is equally murky.

THE TASK OF THE HUMANITIES: LOOKING INTO THE DEEP

The magazine under my feet, as I sit in my office, is an old issue of *Scientific American*. The cover shows an artist's rendition of "The Littlest Human"— genitals carefully obscured—a being that lived thirteen thousand years ago in Indonesia. Other articles are "Quantum Clouds and Atom Chips," "How Ulcer Germs Can Be Good for You," and "Hardwiring Memories into Your Brain." The latter three articles depend upon a feature common to ordinary science: we can examine atoms, germs, and brain tissues using ever more sophisticated devices. We can examine brain tissues using larger and larger machines to see deeper and deeper into neural processing.

Most scientists believe that there are no gaps within nature. Perhaps cancer begins at the quantum level? In 2009, cancer researchers may not have ways to use the work of quantum physicists, but this could change in 2019 or 2029. Whatever causes cells to become cancerous must occur within cells or between cells. It must be a real event in space and time, obeying all the rules of biological, chemical, and physical sciences. To understand that real event, we will invest billions of dollars to magnify cellular processes to a point where we can see how cancer arises. Once we've seen those processes, we can figure out how they work and then devise ways to correct them.

What about the little hairy guy on the cover, the Littlest Human? In contrast to the three other stories, this account begins and ends on a note of confusion. The confusion stems from the word *human being*—or *Homo sapiens*, to use the anthropological term for modern humans. It seems that the Little People were much smaller than *Homo sapiens* and had much smaller brains. If we define *Homo sapiens* by advanced thinking, and brain size determines thinking capacity, then this group seems excluded from the human family. Some anthropologists, though, cite evidence that the Little People made tools; hence, they must have had higher-level brain functioning. Brain size is not everything, they claim.

This question is important because once we ascribe the label *human being* to an entity, it enters into our sense of ourselves. Who we are, what we are, and what we should become are issues that dominate life in the real world. We cannot wait for the National Academy of Sciences, for example, to decide how to raise children. Children are born without directions packaged with them; parents must decide how to raise them without delay. So too, issues of marriage, sexuality, the creation and distribution of wealth, and the use of deadly power press in on us. We cannot wait for a series of studies that, when published, typically are inconclusive.

The task of categorizing the Little People turns on our definition of *Homo sapiens* ("thinking man"), which turns on our definition of the concept *thinking*. Concepts—ideas, images, stories, and accounts—are categories with which we make sense of our world. Ludwig Wittgenstein, a great Austrian-English philosopher, called concepts "tools." This is an adroit, metaphoric, way to explain them. Like a hammer and screwdriver, the concepts of (full-fledged) *human being* and *thinking* do different kinds of work.

The concept *human being* is a tool that, when used by people in authority, determines how a group of beings might live or die. In the eighteenth century, the framers of the U.S. Constitution bent to pressure from Southerners and defined persons of African descent held under slavery as 3/5th a human being. It took a Civil War that cost six hundred thousand lives to redress that injustice. Looking at the twentieth century, we know well that ascribing full human status to one group and not to another makes genocide and other crimes likely. For religious persons, the concept of *human being* also distinguishes us from God, or the gods.

Figure I.1 1908 microscope.

The concept *thinking* is a tool used in a thousand important ways. Among these are the following: it differentiates *Homo sapiens* from hominids; it distinguishes us from animals (over whom we claim dominion); it names the set of skills that we hope universities instill in their students; it occupies the center stage of research in psychology and the tradition of thought we call philosophy. Thinking seems to be that capacity responsible for human evolution and its exponential growth, including the rise of the sciences.

Because concepts like *human being* and *thinking* are so important, we need to understand them as best we can. How shall we do that?

An image of an antique microscope from 1908 sits next to the *Scientific American*.[1] Lovely in its own way, it is now a collector's item esteemed for its finish and design. Later microscopes long passed it by in power and importance to biological research. Like telescopes, their distant cousins, microscopes are tools that let us examine natural objects with more and more exactness. With refinements in lenses, illumination, and numerous other changes, this tool from 1908 has been improved by a factor of many thousands. Can we say the same for studies of the Hebrew Bible?

What kind of microscope would help us decide these issues? None. Are there microscopes or CAT scans we can use to magnify concepts in the way that we can magnify cancer cells? Here again the answer is no. To understand concepts, we need to see them in operation. (To learn how a hammer works we need to see a carpenter use it to drive a nail, a job not suited to a screwdriver.) Disciplines associated with the humanities — such as philosophy, literature, comparative languages, art history, and religion — help us investigate how human beings think about being human.

My distinguished friend James spent his long, productive career struggling with that task. He tried to understand how early Jewish thinkers understood the concept *human being* and how that concept pertains to the concept *God*. James never used a microscope or any other imaging device to carry out his research. Humanists may say that they seek a deeper understanding of Jewish thought, for example, as I seek a deeper understanding of my patients' lives. In both cases, we use metaphors of depth and vision to describe our work.

In this book, I explore these and related metaphors and show how they lead us astray. If the word *deep* means that our subject matter is important, I agree that humanistic subjects are deep. Typically, however, the notion of depth and its numerous offshoots stems from a metaphor of looking in the way that biologists look through microscopes or astronomers look through telescopes. This, I suggest, cannot be true. The objects of humanistic study are central to how we conceive of human beings and their destiny. In order to understand these objects, we must see how they are structured compared to the objects of natural science. We do that in the next chapter.

MAGNIFYING TRUTHS: TWO SLIDE SHOWS

The question of progress in the humanities is an ancient one. To offer a new answer, I suggest we compare the objects of humanistic inquiry to the objects of natural science. They are radically different. To show this, I use an unusual device: the magnifying lens. Using the magnifying lens, we can distinguish two types of object: one type (objects amenable to natural science) can sustain magnification; the other type (objects typical of humanistic inquiry) cannot. I demonstrate this thesis in two slide shows; the short version summarized above, and a longer version entitled "Magnification" on my web page.[1] These slides show us a simple truth: we can investigate the form and structure of natural objects to the very limits of imaging capacity. They can be magnified to any desired size without losing signal: there is always more to see.

The opposite is true of what I've termed cultural objects: they *cannot* be magnified. On the contrary, as evident in the "Magnification" program, objects like photographs, written and spoken languages, and paintings cannot be magnified beyond 2x or 3x without dissolving into mere noise. At that point the signal disappears into a cloud of vague shadows and we can see nothing more.

This simple distinction has profound consequences. First, it means that humanistic inquiry cannot match scientific inquiry in this context. Because the objects of natural science permit magnification, we can dissect them into smaller and smaller pieces, each time discovering new, important information about them. With each advance in magnification (such as microscopes, X-rays, MRI machines, PET scans) comes an advance in knowledge. This advance is an essential feature of what we call science. True, we also wish to know how natural objects work. We observe them *in vivo* or in operation. We cannot understand the kidney merely by dissecting it. Yet, even when we observe the kidney's functions, we do so at multiple levels of inquiry: At the gross, clinical level we observe patients' responses, such as urine output. At each lower level we use modes of magnification to see better how each system operates. Science correctly aims to reduce the mysterious functioning of complex systems like the kidney to smaller, less complex mechanisms whose operation we can understand and control fully.

For good reasons, humanists who produce works of art, such as novelists and painters, and those who love these works of art disdain attempts to "reduce" their subject matters to something smaller than the piece itself. Any criticism that makes the works of Henri Matisse simple and formulaic either is false or destroys what we love about Matisse.

Images like Blue Nude IV, a famous cutout by Matisse, appeared in 1952 when he returned to an apparently childish medium—colored paper and scissors—and created what many feel are his masterpieces. The image works in many subtle ways; none of them will be illuminated by reducing the blue forms to smaller pieces of blue, and the subtle negative space between the right arm and the head to "mere space." As critics have noted, Matisse makes the blue and the white spaces beautiful; good art criticism helps us see this and perhaps another hundred features of these works that we would not otherwise notice.

Another consequence of the difference between natural objects and cultural objects is that large-scale research in the humanities cannot mimic "Big Science." We cannot divide up our subject matter into discrete tasks, assign each to a research team, and rest assured that the finished pieces will fit together when reassembled. In the so-called war on cancer, natural scientists battle over funding priorities and allocations just as physicians battle over market share. But neither group disputes that some research teams must analyze DNA, some must carry out clinical studies with patients, another must do epidemiology, and numerous others must study vastly different phenomena. While conceptual struggles appear in the study of cancer, in principle we can find some rational way to divide up the task of investigating it.

No such unanimity occurs in the humanities. While English departments typically teach Shakespeare, their dominion is more or less arbitrary. A strong theater department, or European Studies program, or history department can challenge their claim to the Bard. New humanistic disciplines, such as Women's Studies and Film Studies, appear in many venues: some emerge within departments, others become autonomous, and still others become divisions of study. This small feature belies a conceptual problem: we cannot say which borders are secure and which are not. Which borders are justified? Which are accidents of tenure awarded to long-lived colleagues?

When humanists talk about "depth" we are not talking about magnification but are using a metaphor. Drawing upon various sources, we are tempted to speak of a *profound* work of art. This means that it occupies a vital place in our sense of our lives. It speaks to us; it moves us; it may help us reorganize our sense of being human. Like psychotherapy, which we consider in chapter 6, art can foment change in the viewer and in the larger culture. Yet, none of these features of great art parallel the notion of hierarchy and depth in the natural sciences. Matisse's wonderful piece is not deep, if that means it can be plumbed by microscopic study. If we magnify the blue paint, we find natural objects that make up the pigment, but we will not find the essence of Matisse's genius or discover why his work is so compelling. To do that, we have to assess the work against other works and assess how the blue and white spaces work together to form the composition. Beyond these aesthetic issues is the important fact that the piece portrays a nude—stylized, true—but still a nude woman, and she evokes another myriad of feelings and associations. Beyond those issues are the historical and contextual features of Matisse's life, the history of European painting, the meaning of Fauvism (his so-called school), and other features of the environment out of which his work emerged.

Throughout this book I take art seriously. I try to show how films, statues, plays, and other forms of art gain their important place in our lives. My readings and comments on these pieces are fragmentary and incomplete; a better critic would help us see more. An even better critic will have something novel to discover about the "meaning" of Matisse for us in our times, a hundred years after he first shocked the establishment in Paris. (As I note below, humanist theorists often call us back to reconsider a work of art of a theory of art, or religion, or philosophy.)

SEARCHING FOR THE HERO: THE ONE WHO KNOWS

That the work of art and other humanist objects, including humanist theories, develop out of individual achievements causes us to focus upon great persons. We do find that social histories of art and museums are filled with great art that is anonymous. Yet, at least since the classical Greeks, we are taken by the story of the individual artist and his or her achievements. This leads us to the individual humanist, the great man or woman, whose mind we find fascinating. By proposing new, exciting ideas, they let us relocate and reassemble our esteemed objects of inquiry. Freud let us reread Shakespeare; feminist critique let us reread Freud; Postmodernist theories let us revise and review feminist theory. Brilliant new ideas and charismatic thinkers give us hope that the treasured past is within our reach: our new way, our new theory will take us back to icons that seemed on the verge of losing all meaning for us. The new theorist seems to have seen more deeply than anyone else. We can see Matisse in a new way. He or she has seen below, into the subterranean depths of the psyche — into the Unconscious (Sigmund Freud); into the Collective Unconscious (Carl Jung); into the very heart of Being (Plato, Martin Heidegger); or into Language itself (Ludwig Wittgenstein).

Not all humanistic enterprises depend upon a great name; yet, at least in European and North American contexts, it is the norm. Psychoanalysts and numerous others locate Sigmund Freud at the center of discussion and debate for more than eighty years. Shortly after Freud's death in 1939, Ernest Jones, his official biographer, compared Freud to the ancient Greeks who proclaimed "Know Thyself!" — the injunction carved over the temple at Delphi. In Jones's estimation, Freud was the first person to obey fully this demand. How Freud did this, when others had tried and failed for over two thousand years, "must remain a cause for wonder. It was the nearest to a miracle that human means can compass, one that surely surpasses even the loftiest intellectual achievements in mathematics and pure science."[2]

As a psychoanalyst, I share Jones's love for Freud. I remember first reading Freud in college. I was shocked by his openness, his zeal for discovery, and his ability to talk about sexual feelings that seemed impossible to put into words. Like Hamlet, Freud talked about the death of fathers, a theme that has dominated my emotional life since I was a boy. I recognize in myself the urge to idealize Freud as the Great Originator, as the Path Breaker. Because Freud took upon himself the task of forging a scientific humanities, I reflect upon psychoanalysis throughout these chapters. I remain a psychoanalyst and remain grateful to Freud, but by the end of the book, I hope that I will have gotten a little farther away from idealizing him.

The problem with idealization is that it makes self-criticism impossible. Where there is no self-criticism there can be no progress. If the great man or great woman has declared that "hysterics suffer mainly from reminiscences" — as Freud did in 1895 — there is nothing more to say, except to elaborate upon those small things that the master has not said. When the master changes the formula, we can follow him into his new discoveries. Like Christopher

Columbus, fame and honor (and dishonor) accrue to the pathfinder who discovered a vast new territory.

That Freud and others can exert so powerful an influence upon us is the puzzle at the heart of this book. The urge to idealize Freud (or Anna Freud or Virginia Woolf) signifies that our fundamental need for order and sureness is not met any other way. Because we cannot count upon a tradition of rational progress, guaranteed by our methods, we must count on something or someone else.

This urge, this instinctlike yearning for someone who knows, does not disappear in persons of genius. We see it reappear in the poignant story of Hanna Arendt and her idealization of Martin Heidegger. Hannah Arendt (1906–1975) entered Marburg University at age eighteen. Having studied classical languages earlier and being intellectually gifted, she quickly became famous on campus as the most brilliant of students. She soon came to the attention of Professor Martin Heidegger, who was composing his masterwork *Being and Time* (1927). Heidegger and Hannah fell in love, and they carried on an intense, secret affair for about two years. For complex reasons, Heidegger did not become Arendt's thesis director. Instead, he sent her to Karl Jaspers, a distinguished philosopher-psychiatrist, at Heidelberg. Under Jasper's tutelage, Arendt wrote her thesis on love in St. Augustine.

When Adolph Hitler became German Chancellor in January 1933, anti-Semitic laws went into effect, oppression of Jews on university campuses increased, and many secular Jews, like Arendt, realized that their very lives were in danger. On March 23, 1933, the Reichstag, the German parliament, passed the Enabling Act ("Law for Removing the Distress of the People and the Reich"). It gave Hitler dictatorial powers and destroyed all forms of civil protection. During these months, Heidegger joined the Nazi Party, sought and accepted the post of rector (president) of Freiberg University. In May 1933, he gave his inaugural address in which he praised the new beginnings made possible by Hitler. That same year Arendt and her husband fled to Paris. In 1941, after much hardship, she escaped to the United States. Based on her firsthand experiences with fascism and her reflections on Europe during the Nazi period, Arendt published *The Origins of Totalitarianism* in 1951.

In the meantime, Heidegger left the office of rector in 1934 (for various, disputed reasons). He remained a member of the Nazi Party until 1945. Heidegger's cozy relationship with Nazism, his membership in the Party, his treatment of Jewish colleagues, and his speech have roused intense debates about his ethics. Critics castigate him for his pursuit of self-interest—including his zeal in making students give the "Heil Hitler" salute—for not defending better his Jewish mentor, Edmund Husserl, and for other failings. (Jaspers, married to a Jewish woman, kept poison nearby should he be arrested by the Gestapo.)

Heidegger's defenders answer these charges with a variety of countercharges and explanations.

Apart from this debate about Heidegger's ethics is the mysterious fact that Arendt did not break off her relationship with him even after his Nazi dalliance. This, superficially, seems unimaginable. Arendt knew well what Nazi politics meant in action; she must have had many near-death experiences while fleeing Nazis intent upon murdering her and her husband. She certainly knew the extent of the Nazi war against the Jews and wrote about it at length in her controversial study of Adolph Eichmann in *Eichmann in Jerusalem: A Report on the Banality of Evil* (1963). While the fact of her affair with Heidegger was not known until much later, that seems insufficient to account for Arendt's continued admiration of him.

That undiminished affection appears in a famous article Arendt wrote about Heidegger for his eightieth birthday.[3] There we learn that her attraction went far beyond the common professor-student romance. Heidegger offered Arendt and other students a vision of progress. With a novelist's skill, Arendt describes how they had heard rumors of a hidden king whose lecture notes circulated everywhere. Because Heidegger was able to link her and other students to the beginnings of philosophy (the Greeks), Arendt says, his beginning had something divine about it: "If it is true, as Plato once remarked, that 'the beginning is also a god; so long as he dwells among men, he saves all things' (*Laws* 775), then the beginning in Heidegger's case is . . . the first lecture courses and seminars which he held as a mere *Privatdozent* (instructor) and assistant to Husserl at the University of Freiberg in 1919."

Although he lectured upon ancient writers, Heidegger was no mere professor, no classicist who dissected ideas *about* philosophy. He *did* philosophy. By the early 1920s Heidegger was teaching himself and his students how to *think*, how to take part in the libidinal process of philosophic reflection. Heidegger, Arendt says, was uncovering truths hidden since ancient times:

> There was someone who was actually attaining "the things" that Husserl had proclaimed, someone who knew that these things were not academic matters but the concerns of thinking men—concerns not just of yesterday and today but from time immemorial—and who, precisely because he knew that the thread of tradition was broken, was discovering the past anew.

This 58-word sentence—translated by Albert Hofstadter from Arendt's German original—is invocation and refrain. "From time immemorial" refers to our wish to feel that someone greater than we is connected to the beginnings, to primordial and eternal truths. He (or she) is a living bridge between us and the transcendent moment of the beginning. Catholic dogma holds that every new pope is linked to all previous popes, who, in turn, are linked to the Apostle Peter, and Peter to Jesus who blessed him and founded the church upon him. Heidegger had found what was lost, he had retrieved the broken thread of the sacred tradition.

Like others who describe the origins of the hero, Arendt uses rhetorical devices to frame our expectations. Arendt's sentence has five refrains. Each heralds a Great Man:

> *someone who attained "the things" that Husserl sought*
> *someone who knew these things were not academic*
> *someone who knew the concerns of thinking men*
> *someone who knew that the thread of tradition was broken*
> *someone who knew how to rediscover the past*

Given the state of German universities in 1924, the destruction wrought by the Great War, the collapse of civil government, the threat of revolution, and public crises that swept through Germany, it is not surprising that people yearned for stability and permanence. Arendt and other thoughtful commentators upon this epoch have described these contexts in wrenching and beautiful detail. I do not dispute them. I merely note that this yearning for certainty made Heidegger's answer, rich with subtlety and newfound concepts, irresistible.

Tied to this philosophic answer was Heidegger the man. It is he whom Arendt defends in her article. We note how unlike this is in the natural sciences or in the social sciences: Newton the man seems unimportant to Newtonism, just as Darwin the man seems unimportant to the theory of natural selection. In contrast, when Ernest Jones praised Freud for having seen deeper than anyone since the Greeks, we hear the tone of idealization of Freud the man, not Freud's thought. As Arendt notes, ideas and philosophical systems are doubtable and refutable: "For it is not Heidegger's philosophy, whose existence we can rightfully question (as Jean Beaufret has done), but Heidegger's thinking that has shared so decisively in determining the spiritual physiognomy of this century."

Heidegger's thinking is not mere reasoning or problem solving; it is not a technique that can be formulated. No; again and again Arendt, like Heidegger, rejects the very idea of elucidating the nature of this kind of thinking. And like Ernest Jones speaking of Freud's genius, they both tend toward an occult claim that thinking of this type occurs within the self, in parts of the self that are unknown and unknowable. Using this metaphor, Arendt claims that Heidegger alone made philosophy possible in the twentieth century: "That something like Heidegger's passionate thinking exists is indeed, as we can recognize afterward, a condition of the possibility of there being any philosophy at all. But it is more than questionable, especially in our century, that we would ever have discovered this without the existence of Heidegger's thinking. This passionate thinking . . . can no more have a final goal—cognition or knowledge—than can life itself."

This sentence cannot be valid. For it seems that Arendt is saying that Heidegger rediscovered a kind of philosophic passion, a true form of wonderment first seen in the Greeks, and this wonderment is essential to doing philosophy. If so, why could others not have rediscovered the same thing? Or, since neither we nor Arendt know fully the history of thought, are there not other writers who make precisely the same point? Finally, Arendt's sentence is logically suspect. We can translate it at least two ways. One is: "In the entire twentieth century, only Heidegger preserved true philosophy." This is grandiose but not illogical. Another way to translate Arendt's sentence is: "If Heidegger had not lived or written then we would have lost this link to the past." This sentence is a problematic conditional: since Heidegger did live and write in the twentieth century, talking about what would have happened if that had *not* occurred puts us in the murky world of counterfactuals.

These questions may seem silly compared to the force of Arendt's prose. Reading her essay is like hearing a brilliant summation by defense counsel. Of Heidegger's job as rector, she says that he succumbed to temptation and entered human affairs: "He was still young enough to learn from the shock of the collision, which after *ten short hectic months* thirty-seven years ago drove him back to his residence" (emphasis added).

Arendt's sentence is not focused enough to be refutable. Its effect is to make Heidegger an absolutely irreplaceable person whom others tempted into error. He alone rescued philosophy and the very possibility of doing philosophy for us, his heirs. Given his place in Arendt's heart and her core identity, Heidegger's error for those ten short months (when he was forty-two years old) is of minor concern.[4] From the standpoint of the vast reaches of time—looking backward to the Greeks and forward to generations of philosophers who will follow—ten months is trivial: "The wind that blows through Heidegger's thinking—like that which still sweeps toward us after thousands of years from the work of Plato—does not spring from the century he happens to live in. It comes from the primeval, and what it leaves behind is something perfect, something which, like everything perfect (in Rilke's words), falls back to where it came from."

By summoning up Rainer Maria Rilke (1875–1926), Arendt links Heidegger to a great German poet and to Rilke's poems about primeval nature, which, like the wind, is beyond good and evil. It is unusual to find a rational person speak about "something perfect" in human nature. Arendt does, however, and in so doing she illuminates the religious hold that Heidegger had on her. The "wind" blowing through Rilke's wonderful poems is individualized, creative experience.

LARGE-SCALE RESEARCH IN THE HUMANITIES

Interdisciplinarity is a feature of large-scale research in many fields and is the hallmark of Big Science. Ambitious projects in contemporary medicine and natural science often require teams of hundreds of researchers. Alongside these examples, a small research team, say in botany, can draw upon a scientific literature created by thousands of authors beginning in the seventeenth century. While no one would assess each of those articles equally, we value most of them with some degree of assurance. We assume that contemporary botanists need not repeat every experiment recorded in the botanical literature. "Progress" denotes our belief that a scientific literature contains enough valid observations that we can go forward and depend upon it most of the time. We anticipate and find a volume named *Progress in Botany*. It is designed to keep "scientists and advanced students informed of the latest developments and results in all areas of the plant sciences."[5]

No such anticipation dominates disciplines like philosophy, literature, and theology. We do find volumes with names like *Progress in Theology* or *Progress in Literary Criticism* but these are products of a particular school, not summations of progress of the entire field. In fact, we commonly find humanists who ask us to go backward. They ask us to reject claims of progress made in the past two hundred to four hundred years. Esteemed figures like Martin Heidegger, Carl Jung, and Jacques Lacan became famous and influential by asking us to return to an earlier era. Lacan pushes us back to Freud (some one hundred years); Jung pushes us back to theologians of the Middle Ages (five hundred years); Heidegger pushes us back to the ancient Greek philosophers (about twenty-five hundred years ago). We could add to this list names like Michel Foucault, who rejects the past two hundred years of psychiatry, and Alasdair MacIntyre, who rejects the past three hundred years of ethical theory and philosophy.

It is hard to imagine this pattern occurring in any field labeled "science." This is not to say that these calls to step backward are regressive or automatically signs of a conservative agenda. On the contrary, Foucault and MacIntyre have helped recast their fields by mounting rigorous critiques of the claims of progress. This illustrates the paradox of progress in the humanities: those whom we esteem the most and around whom schools of thought develop are often thinkers who reject claims of advance. With many brilliant asides, Foucault exposes fault lines in schools of psychiatry that hoped to ape the natural sciences. MacIntyre's critique of ethical theory, in his magisterial study *After Virtue* (1984), calls into question much of the self-understanding of Modernism in philosophy and politics.[6]

Thanks to a generous university, I serve as director of the Center for the Study of Religion and Culture (CSRC) at Vanderbilt University. Part of the CSRC's mission is to bring together teams of scholars from as many of Vanderbilt's ten schools as possible. I've enjoyed some four years of discussions with about sixty colleagues at Vanderbilt University on this question: how can we promote large-scale research in the humanities? This is another way to ask: how do the humanities as taught in American universities, at least, differ from the sciences and the social sciences?

Seven models may be used to describe humanistic research.

20-MULE TEAM

While charming in the way that only mules can be, this metaphor collapses differences into unity. It presupposes that we're all the same kind of animal, pulling in the same direction, with the same mindset and goal. It also assumes that there is one driver—the mule skinner—who knows what the team should do and where it is going.

CHOIR

This model captures the diversity of "voices" and skill sets that we hope to see in large-scale research projects. Like the jazz band, orchestra, and theater troupe, it evokes the possibility of making good music together, of an eventual harmony in which everyone has a proper place. Yet, it presupposes a conductor or choir director and it presupposes a single piece of music. The choir must literally be on the same page, following a score already written and completed. Choirs do not create new music; they perform something produced prior to their performance. None of this is true of CSRC projects or most projects in combined social-science and humanistic research.

SPORTS TEAM

Perhaps humanities projects should be organized like teams, say the Iowa State baseball team of 1888? If CSRC projects are analogous to teams, then we would all have assigned places; strict rules; a way to keep score; and measures of progress, namely, winning and losing in struggles with other teams. As any baseball fan will tell us—even without asking—endless debates ensue over quantitative questions and ranking. Is Babe Ruth greater than modern ball players since he played in an era deficient in health care, steroid uses, and other benefits of the modern game? Few of these features of a sports team, however, match the typical humanities project. We seek the right combination—for example, a natural scientist and a sociologist—but the grounds for our search are subject to change depending upon a host of accidental features: who's on sabbatical, who gets on well with others, who is willing to join us? Most telling is the absence of clear and reliable metrics—we don't have home runs, RBIs, and similar measures essential to the expert's appreciation of the immortal game.

LIFEBOAT

As illustrated in Hitchcock's great film *Lifeboat* (1944), a lifeboat contains a random collection of survivors. The passengers include a millionaire, a colored porter, a society lady, a workingman, and so on. Depending upon their luck and fortitude, they will be rescued or left to an unhappy fate. Part of the drama is the emergence of leaders. Typically, a crew member from the ship fails to do his duty, and a civilian must step forth. As part of the war effort, Hitchcock amplified this theme into a call for civilian cooperation and sacrifice. It also let him delineate the nature of Nazi cruelty. After the lifeboat passengers rescue a crew member from the U-boat that sank their ship, he repays their kindness with betrayal and murder. The German seems invincible until the otherwise peaceful civilians mobilize themselves to destroy him, just as England, the United States, and their allies rose up to destroy Nazi Germany.

This semidemocratic process seems ideal, in this circumstance, for a leader emerges during these moments of terror. The story of the lifeboat makes for exciting cinema. (Hitchcock asked Ernest Hemingway to write the script; when he declined, Hitchcock brought in John Steinbeck.) But no one would wish to use this as a model of intellectual group behavior and organization.

DISTRIBUTED COMPUTING

Distributed computing designates a network of computers linked to one another, running parts of a shared program, and designed to handle large tasks. In a minimal sense, numerous websites that permit exchange between users are instances of distributed computing. But in a more rigorous sense, distributed computing involves "breaking down an application into individual computing *agents* that can be distributed on a network of computers, yet still work together to do cooperative tasks."[7] An agent is a well-defined component of the system assigned a particular function. In complex banking systems, for example, the software and hardware that interact with customers at the ATM interface form an agent.

A popular instance of distributed computing is the network called SETI@ home. It links thousands of personal computers (in homes, offices, universities, etc.) to search for extraterrestrial intelligence, "a scientific experiment that uses Internet-connected computers in the Search for Extraterrestrial Intelligence (SETI)." Anyone "can participate by running a free program that downloads and analyzes radio telescope data."[8]

Unlike movies, images of a typical distributed computing networks are boring. Certainly they lack the drama of *Lifeboat*, the charm of the mules, and the emotional reach of the children's choir. Instead, the beauty of distributed computing lies in its output, the solution offered to its users. If SETI@home helps us discover the existence of extraterrestrial intelligence, it will be part of a momentous event in human history. If SETI finds genuine signals from nonterrestrial intelligences, it succeeds. If a bank saves money on transaction costs and improves service to millions of people hovering over its ATMs, it succeeds. If the National Hurricane Center is able to make accurate forecasts of hurricanes, it succeeds.

The metaphor of distributed computing is attractive because it suggests an image of dozens or hundreds of humanists integrated and working together, assembling their intelligences toward a common goal. The metaphor breaks down quickly though because distributed computing includes at least four elements rarely available to humanists: (1) a common definition of the shared problem; (2) a problem that can be reduced to equations or computational tasks; (3) an overarching program that sets out the precise rules for sharing the computational burden; and (4) a clear, shared picture of what will count as a solution.

BIG SCIENCE

Universities and national research labs are familiar with large-scale research projects with teams of researchers spread out around the world.

A recent picture on the cover of *Physics Today* shows a view of a massive experiment named PHENIX running at Brookhaven National Labs in Long

Island, New York. The largest of four experiments using the Relativistic Heavy Ion Collider (RHIC), it is designed to investigate what happens when you smash together heavy ions at near-light speed: "The primary goal of PHENIX is to discover and study a new state of matter called the Quark-Gluon Plasma."

Why spend millions of dollars over many years to study minute interactions lasting a very tiny amount of time? Because our best, current theory of the strong (nuclear) force, the fundamental force that binds quarks into protons and neutrons, is notoriously difficult to use. This new state of matter is one of the rare testable predictions of this theory. Science tells us that the universe began about thirteen billion years ago in a specific event called the Big Bang.[9] About ten millionths of a second after this event, the expanding universe was composed of unbound quarks and gluons before they coalesced into more complex particles that eventually became galaxies, planets, and people. To put it more dramatically, PHENIX aims to look into the state of matter, the quark-gluon plasma, that existed just after it all started. The RHIC website describes the project for laypeople: "RHIC will recreate (on a small scale) the temperatures that existed at the dawn of the universe."[10]

While it is a notable instance of Big Science, PHENIX is not unusual in its history, structure, and function. On the contrary, we find many instances of large-scale projects in the physical and biological sciences. For example, mapping the human genome, a task begun in 1990 and completed in 2003, took immense effort, money, and skill. No comparable projects appear in the humanities because humanists cannot rely upon a research tradition in which fundamental questions are answered, leading to new questions and thus a progressive tradition.

Common to massive science projects are four features: (1) a unified theory that is clear enough and persuasive enough to gain unanimous approval (even if temporary and given grudgingly); (2) a lengthy period of gestation of key ideas, experimental evidence, funding, and experimental expertise; (3) a master plan or blueprint that lays out the sequence of investigation and research; and (4) a systematic way to assess the validity of findings. The last is essential to any science, one might say, but in these large efforts the burden and anxiety are greater and thus the need to convince governments and other funding agencies is greater.

I am not aware of any large-scale humanities research project that shares these four features. The closest example is the *Cambridge Dictionary of Christianity*, edited by my colleague Daniel Patte. The work is lengthy and involves hundreds of scholars. However, they labor as individuals: each writes a substantial article for the dictionary according to the dictates and values of his or her academic specialty. Daniel's work is more novel because he and his coeditors must conceive of an overarching schema for the dictionary. Thus, while the editors must struggle with providing a more or less unified vision for the dictionary's hundreds of articles, the individual authors need not. The dominant structure of the dictionary is (correctly) horizontal—its mission is to describe the diverse forms of Christianity around the world. Though daunting and essential, this task offers little in the way of a unitary or explanatory theory of Christianity. True, within selected articles one may find a variety of theories of religion (such as sociological, psychological, literary, feminist). The plurality of these theories and their lack of common definitions of what counts as "Christian" illustrate typical features of humanistic discourse.

It may appear that the discussion above is nothing more than a series of negative judgments, a kind of *via negativa*. We can say what humanistic working groups are *not*. They (and, by extension, similar projects in hermeneutic-social science research groups) are not structured precisely like any of these six models. Here are some things that humanistic groups typically lack:

1. a common definition of a shared problem
2. a unified theory (or script or master score that locates everyone)
3. an overarching program that sets out the precise rules for sharing the computational or research burden
4. measures of progress
5. a shared vision of what success would look like

Looking more closely at this list one can condense these five items into one: members of large-scale humanistic research teams often lack a shared agreement as to what we're talking about. We have general notions of what we mean by "religion" or "the text" or "Christianity," but once we bring these subjects to a more exact accounting, troubles arise.

To put it simply, we don't begin our work with shared definitions. This truism stems from a more interesting feature, which is that we don't have a common sense of our *objects*—that is, the things that we're investigating. Lacking a shared point of origins, we cannot proceed to agree upon a common, overarching theory, nor to a shared metric, nor to generally accepted measures of what counts as progress in our investigations.

This does not mean that humanistic inquiry is fatuous, or that its objects are trivial. On the contrary, humanists and hermeneutic social scientists talk about the essence of personal and public self-understandings. Momentous issues of identity, justice, war, and peace turn on these forms of self-understanding. These are religious questions because they turn on how we think about the nature and destiny of human beings, as Reinhold Niebuhr put it.[11] Recent debates about end-of-life treatment, procreation, the Gulf War of 2003, the rise of orthodox and ultraorthodox sects, the future of free trade, the distribution of wealth, and the threat of global warming will not yield to ordinary science, even to teams of scientists. While these questions are immensely important, they are also immensely diffuse. They are larger than the scope of a single discipline or single theory.

SKUNK WORKS: DISCOVERY AT THE EDGES

An image much closer to the work of humanistic groups is a union of European salons and what Americans call "skunk works." For example, the famous salon of Henriette Herz (1764–1847), a Berlin salonière (or "muse"), may be compared to the work of the team that created the F-117, the U.S. Stealth Fighter, designed by a team at Lockheed Corporation that was known as the "skunk works."

Women like Henriette brought together extraordinary talents from the arts, sciences, professions, and politics who would not otherwise have known each other.[12] Like Hebe, the Greek goddess of youth and the cupbearer to the Olympian gods, Henriette is pictured as a beautiful woman, with springtime boughs in her hair, a golden cup of wine, and a direct, thoughtful and focused gaze.[13] As Henry David Thoreau (1817–1862) declared, Hebe stands for a restorative vigor, a goddess of possibilities: "I am no worshipper of Hygia, who was the daughter of that old herb-doctor Asclepius . . . but rather of Hebe . . . who had the power of restoring gods and men to the vigour of youth. She was probably the only thoroughly sound-conditioned, healthy, and robust young lady that ever walked the globe, and whenever she came it was spring."[14] (Her name comes from the Greek verb *hēbaō*, which means to be in the prime of youth; associated meanings are to be passionate, prime, fresh, and vigorous.) Denied entrance to the professions, Jewish intellectual women founded salons in which the art of conversation, challenge, and discourse flourished.

While sharing these values, humanistic research groups have an intentionality foreign to the salon. In this sense, humanistic groups are analogous to skunk works, groups like those at Lockheed who are given vague, but essential tasks to develop a new kind of fighter (the first American jet), spy plane (the U2), stealth technology, and other challenges, some of which, of course, fail.[15] While their budgets are in the tens of millions of dollars and ours are in the tens of thousands, we have in common the task of organizing a group of disparate persons, with vastly different training and backgrounds, to confront something never done before. Numerous accounts of American corporate breakthroughs, like the IBM personal computer and other innovations in computing, came out of similar skunk work groups.[16] Another example is the Palo Alto Research Center

(PARC) of Xerox Corporation begun in 1970. It is credited with laser printing, Ethernet, the graphical user interface, and "ubiquitous computing."

Like members of a salon, humanists talk across boundaries and sometimes head toward conceptual cliffs (or abysses). Like members of skunk works, our groups are also directed toward rethinking accepted wisdom and the restraints of the disciplines. This means that there must be many false starts and many tumbles along the way. What each does not know are the upper and lower limits of his or her discipline. Unless one has ventured beyond the confines of a discipline—which we are taught to never, never do on the way to tenure—one cannot experience the giddy, slightly sickening feeling of being out of one's depth. Our wager is that we can make new discovery at the edges of what we know and what we do not know.

There are many ways to not know; we focus upon the experience of coming across a barrier in one's conceptual world. Rather, like fifteenth-century European sailors fearful they might sail off the edge of the world if they ventured too far, we can fear the same. It is also the tradition of normal science, as T. S. Kuhn described it: a rich, not-yet-exhausted paradigm generates a series of scientific puzzles, and to their solution many generations are correctly devoted.[17] Only when the number of anomalies increases and the number of solutions decreases does normal science enter into a crisis. This description of normal science does not fit the world of religious studies, much less the questions that dominate humanistic discourse.

SELF-UNDERSTANDING AS THE OBJECT OF HUMANISTIC RESEARCH

Again, why is progress so difficult in the humanities compared to those disciplines aligned with the sciences? The short answer is that the object of humanistic inquiry is human self-understanding. Human self-understanding is organized along horizontal, not vertical axes. This means that we cannot dissect, for example, Islamic theology, taking "slices" of it the way we can dissect and understand a natural object.

Islamic theology organizes the thoughts and feelings of a billion people. Catholic theology does the same for another billion people; we cannot understand either without entering into the myriad horizontal contexts of family, tradition, language, and the maturation of self-understanding, with all its mysteries, that coalesce over a lifetime into the identity "Muslim" or "Catholic." These two terms name something of profound importance. In complex ways, thoughtful Americans have struggled to understand the attacks on September 11, 2001, and to locate those reasons within the larger context of Islamic thought. Immediately following the attacks, President Bush and others conceived of the conflict as part of the "global war on terror." Careful to not blame all Muslims, the American administration has had to confront the fact that those who designed and carried out these attacks saw them as religious duties. Initially, the war on terror was carried out against those agents, especially those organizing within Afghanistan. To battle them, the United States conducted a traditional military campaign and brought it to a brilliant and successful conclusion, as it did in the second Gulf War. According to the usual categories, then, the United States was wining the war that President Bush had declared in the fall of 2001.

This changed in July 2005. The U.S. government would now call the war "a global struggle against violent extremism."[18] There are no doubt many reasons for this change, but among them is the conceptual problem that the word *war* brings to mind World War II, which the Allies won, and perhaps the Viet Nam War, which the Americans lost. In both cases, wars are conflicts between states that have borders, histories, delegates, governments, uniformed armies, and other structures. Modern wars do not go on forever, but struggles may. We struggle to cure cancer and we struggle to end poverty; no one claims to know exactly when and if these struggles will end. This change of name has all the hallmarks of a humanistic problem. It asks us to understand a perplexing event that touches the lives of millions of people and that deals with human actors at their most extreme.

In times of war, governments employ moralizing categories. Thus, a few years ago, Donald Rumsfeld, the Secretary of Defense, "described America's efforts as it 'wages the global struggle against the enemies of freedom, the enemies of civilization.'"[19] This dramatic statement may appeal to domestic listeners because it aligns us with freedom and civilization, but it cannot portray accurately how terrorist bombers understand themselves. Only cartoon bad guys organize themselves to destroy what is good and crush freedom. In the real world, people willing to die and kill must tell themselves stories in which they, not their victims, are struggling to preserve their own civilization and freedom.

Mr. Rumsfeld was no doubt doing his best to lead the Defense Department; we must hope, though, that away from television cameras, he was also trying to understand the mind of his adversaries. Like most important humanistic matters, this is a task that cannot be postponed. Humanistic problems like that confronting Mr. Rumsfeld arise in specific contexts. We cannot wait a generation or two to respond to attacks like those of September 11, 2001. To go to the other end of the spectrum, a five-year-old girl living with an alcoholic father cannot wait fifteen or twenty years to grow up. She must grow up now, in the fog of terror, struggle, and remorse. Both great nations and little children need to survive day to day, and to survive they must offer themselves an accounting, a story of who they are now and who they might become. Rumsfeld's story, though, is one of good versus evil, as is the young child's account of her suffering. In both stories, the victim feels hated for what she or he is. Most importantly, the sense of historical time is deformed into an ever-present "now" of anxiety and danger. Questions about how we got here, about the social and political realities that shaped the conflict, cannot be addressed. Political actions by the other are reduced to cartoonlike sketches of sadistic intent to destroy what is good. Like an inversion of the Eternal Now of religious experience, these convulsive experiences of dread make it impossible to see oneself in normal time. The past is irrelevant, the future is in jeopardy; everything is changed.

In sharp contrast, Abraham Lincoln's speeches about the U.S. Civil War do not reveal this split worldview. For many Americans, Lincoln's account of the Civil War is an essential part of our self-understanding. A handful of these speeches have shaped American's sense of themselves.

In his address at the Cooper Union in New York City, on February 27, 1860, before he received the Republican nomination for president, Lincoln addressed a historical claim by his longtime opponent, Stephen Douglas. Douglas had argued that because the founders of the Republic had countenanced slavery and, indeed, had recognized it in the Constitution, we, lesser than they, must acknowledge that they had settled the question of slavery once and for all.

Lincoln challenged this interpretation of the history of the Constitution and the meaning each of the founders assigned to it. He spent some 7,700 words and almost two hours showing that, contra Douglas, the founders intended to see slavery curtailed and eventually eliminated:

> The facts with which I shall deal this evening are mainly old and familiar; nor is there anything new in the general use I shall make of them. If there shall be any novelty, it will be in the mode of presenting the facts, and the inferences and observations following that presentation.
>
> In his speech last autumn, at Columbus, Ohio, as reported in "The New-York Times," Senator Douglas said: "Our fathers, when they framed the Government under which we live, understood this question just as well, and even better, than we do now."
>
> I fully indorse this, and I adopt it as a text for this discourse. I so adopt it because it furnishes a precise and an agreed starting point for a discussion between Republicans and that wing of the Democracy headed by Senator Douglas. It simply leaves the inquiry: "What was the understanding those fathers had of the question mentioned?"[20]

The Constitution is a complex artifact, shaped by numerous forces and motives, many of them available neither to the authors nor to us. The role and nature of slavery within U.S. history is yet another complex problem equally shaped by overt and covert forces. The idealized status given to the "fathers," the thirty-nine men who signed the Constitution, is also a complex narrative. Finally, "the understanding those fathers had of the question mentioned" is yet another complex artifact that Lincoln hopes to illuminate in his lengthy analysis.

Narratives about narratives are humanistic objects and can be understood best by humanistic means of inquiry. Humanistic objects are complex, but are they deep in the sense layered in hierarchies that we can examine through magnification? No. To imagine them as deep is to fall into a paranoid worldview: that we can find the hidden meaning of things by examining them in greater and greater detail. We cannot isolate humanistic objects, such as paintings and poems, and drill into them, looking for their essences. The urge to do so is powerful, and many a best seller addresses it directly by concocting a totalistic story in which everything that is puzzling or incomplete finds its proper place in a hidden scheme:

> While in Paris on business, Harvard symbologist Robert Langdon receives an urgent late-night phone call: the elderly curator of the Louvre has been murdered inside the museum. Near the body, police have found a baffling cipher. Solving the enigmatic riddle, Langdon is stunned to discover it leads to a trail of clues hidden in the works of Da Vinci . . . clues visible for all to see . . . and yet ingeniously disguised by the painter.[21]

Like his forebears, especially Arthur Conan Doyle, Dan Brown follows an almost irresistible recipe in which odd parts of Western history, its most illustrious citizen (Jesus Christ), its oldest and most secretive power (the Catholic Church), and its idealized "genius of geniuses" (Leonardo Da Vinci) come together in a story of evil versus good. This is thrilling and vastly comforting. For while the story involves intrigue, danger, and cleverness, it is an understandable story: there is a key, a cipher that a clever-enough hero can grasp and thus solve. This gives the illusion of depth, but depth here refers to hidden actors and hidden motivations. Once the author reveals these hidden motives, the illusion of depth disappears into a fully readable and trite message.

Looking back to Lincoln and his hallowed place in American self-consciousness, we find something quite different from *The Da Vinci Code*. Because Lincoln took upon himself a prophetic stance that altered American self-understanding, Americans are entangled in narratives using concepts and self-understandings derived in part from Lincoln's words—and his self-reflection. Humanistic means of inquiry such as literary criticism, psychology, history, and other disciplines give us ways to construe and reexamine the meaning of these narratives. Because Lincoln's words are essential to American self-understanding, when Americans face new, challenging contexts, we go back to Lincoln's speeches. As long as an American nation exists, there will be new books about Lincoln and other heroes, just as religious traditions must find new ways to reexamine the meaning of their exemplars.

Within specific contexts, this need to know what cannot be known generates narratives, and narratives have their own internal logic. Central to narrative logic is the need to tell a story that explains one's action. The explanation need not be valid; indeed, often we have no way to assess the validity of one story compared to another. However, it must be believable. What counts as believable varies from audience to audience and from context to context. In traditional societies, one might ascribe bad things to the effects of witchcraft; this will not work in modern contexts. In modern contexts, a plausible narrative must include an assessment of motives, feelings, and a self-conscious reflection upon one's past. When done well, a narrative like this can feel persuasive—as long as one remains under its spell.

Because the objects of humanistic reflection are individual selves situated within larger contexts of family, group, nation, and culture, denoting their boundary is not easy. Arendt's defense of Heidegger the Man and Heidegger the Thinker occurs within a matrix defined by her special relationship to him, by his stature as a preeminent thinker, by her feelings about German culture, and by the social context of early-twentieth-century Europe facing decline while the United States rose in power, decade by decade.

Was Heidegger an accidental Nazi? Did his affinity for Nazi ideology resonate with that of numerous, less talented, German intellectuals who also yearned for a total, revolutionary answer to German's malaise following World War I? (When Arendt began university, the savings of middle-class Germans were ravaged by uncontrolled inflation: "In Munich in early November 1923 a light lunch cost well over a billion marks.")[22] Does Heidegger's disdain for democratic value and science—and his idealization of the Greek—prefigure his affection for German fascism? Can we rescue Heidegger's insights from his fascist longings?

These are complex questions addressed by serious scholars. Yet, none will yield to a microscopic treatment of Heidegger's text—or of his dreams of German rebirth. While both of these are essential topics of inquiry, neither is sufficient because neither helps us locate Heidegger in horizontal comparisons against similar thinkers and movements. One might spend a lifetime studying *Being and Time*; one might do the same studying Freud's greatest book, *The Interpretation of Dreams* (1900). Yet, even resolute textual scholars will have to immerse themselves in the thought-world of each author. This will inevitably turn into a comparative study of how each book emerged from within an intellectual ecology, defined by social and political forces. Humanistic research cannot, in this sense, avoid moving from depth studies to horizontal studies, from textual studies to social, cultural, and political studies.

DEEP LANGUAGE: THE ANXIETY OF TRANSLATION

Heidegger says that there are two great philosophic languages: classical Greek and German. This cannot be true. Descartes wrote in Latin and French; David Hume in English; a score of Hindu mathematicians and linguists wrote in Sanskrit; Arab and Jewish philosophers wrote in their respective languages. Heidegger's claim tells us that he was raised speaking German: he came into consciousness through German, and so German has a transcendental aura for him. It also tells us that like numerous others in the West, Heidegger idealized the archaic Greeks. Given his poetic talent, it seems true that Heidegger could not have written his masterpieces in English or French. (Surely masterpieces can be written in English or French or Korean or Afrikaans?)

Further, it may be that to appreciate Heidegger one must read him in German with fluency similar to that of Arendt's. This makes his works similar to other humanistic artifacts—as opposed to scientific treatises—that cannot be translated without doing them violence. To translate a German paragraph about "Sein" into an English paragraph about "Being" is to risk going wrong in numerous ways. It is as hard as translating a beloved poem by Rilke into English. A line in "First Elegies," from Rilke's *The Duino Elegies*, is "Und das Totsein ist mühsam," which means literally, "And being dead is hard work." Nicole Krauss compares two translations: Edward Snow writes, "And death demands labor." J. B. Leishman translates the line as "And it's hard, being dead." Which is the better translation? How could we settle upon a universally shared judgment?

Here we glimpse an anxiety that permeates humanistic reflection. To this simple question of best translation, there is no simple answer. Each answer is a compromise of contending values, and each value derives from specific contexts and needs of translator and audience. Krauss prefers Leishman's version and its "rhythmic, resonant lines."[23] I do too; but I'm immersed in American values, trained by Anglophiles, and admiring of direct, powerful speech. Another person, immersed in different values, could argue that Snow's line, which attributes to death a demand, is better.

Anxiety about the future of one's language and the sense that it is rapidly falling apart did not begin in the twentieth century. Samuel Johnson (1709–1784) reflects exactly this concern about English in "The Plan of an English Dictionary" (1747). Writing to the Earl of Chesterfield, "One of his Majesty's principal

Secretaries of State," Johnson says, "This, my Lord, is my idea of an English dictionary; a dictionary by which the pronunciation of our language may be fixed, and its attainment facilitated; by which its purity may be preserved, its use ascertained, and its duration lengthened." Recognizing that he cannot claim to have exhausted the subject matter, Johnson employs a clever military metaphor to describe his wishes. Like the soldiers of Caesar, he looks upon Britain as a country to be invaded and domesticated: "But I hope, that though I should not complete the conquest, I shall, at least, discover the coast, civilize part of the inhabitants, and make it easy for some other adventurer to proceed further, to reduce them wholly to subjection, and settle them under laws."[24]

The anxiety of translation is that one is never quite right—to translate is to engage in treason. As the Italians say with more élan, "Traduttore, traditore." With vast self-consciousness, Samuel Beckett and James Joyce, to name two great Irish writers, play with translations, each exploring the paradoxes of language. In Beckett's case, he presents the odd example of an esteemed author translating his work from one language to another. In Joyce's case, we find an author whose ear for language and its multiform uses—from the Mass to tawdry advertisements—permeates his comic work *Ulysses*. The mystery, however, and the anxiety of translation appear clearest in the case of poetry and other forms of language that use its accidental features, its sounds, weight, speed, to musical effect. Machines can translate direct, scientific speech without loss. As anyone who has studied a language will attest, one can read a scientific article in French, say, with considerably more ease than one can read a French newspaper. The former is written to express propositions with as little nuance and color as possible; all should be on the surface, with clear and focused meanings. The latter depends upon nuance, suggestion, and ambiguity. A machine does not know how to savor the sound of French vowels but French readers do. French poetry depends upon that and similar human capacities.

Among humanist achievements are dictionaries like that of Dr. Johnson and other works that study language in its myriad of uses. How one says something, which words one uses, which narratives are entangled with them, matter. Like the word *war*, terms like *Christian, science, truth, good,* and *holy* carry with them implications for action. Drawing in part on words from Lincoln's great speeches about government for the people, Secretary Rumsfeld urged Americans to understand the war in Iraq as the struggle for democracy. If that were true, if we all agreed that the war in Iraq is analogous to the American Civil War and that its objective is to expand human freedom, then the debate would melt away. However, Rumsfeld's speeches, like those of George W. Bush, are closer to *The Da Vinci Code* than they are to Lincoln's Second Inaugural.

As Arendt said of Martin Heidegger, Rumsfeld's words are not mere cognitions or statements of fact; they cannot be refuted by pointing to mundane facts on the ground. Rumsfeld, it appears, grounds his speeches upon a distinct view of military power and its ability to shape the future. Not knowing that future, we cannot assess his speech, except to contrast it to another view of war and, perhaps, another philosophy of human being.

This brings us to yet another anxiety that humanists must confront: our objects are shaped by our beliefs. Apart from the fascinating but quite limited ways that natural scientists must deal with quantum effects, the objects of natural science do not vary whilst under the microscope.[25] The opposite is true of human beings viewed as individuals and as collectives. As parents and psychotherapists know, the theories of personhood that we bring to raising children or responding to psychological suffering affect directly how the other person responds to us. At collective levels, a totalitarian theory of government, which mimics paranoia, will create the very conditions of suspicion and counterrevolution most feared by those in power.

Living in Germany after World War I, Heidegger and Arendt knew a kind of disintegration anxiety that most Americans cannot grasp. Even during the Civil War, Southerners knew that, should they lose, Lincoln would not annihilate them. Middle-class Germans who watched their life savings melt away in economic collapse could look at Russia, not that far away, and see there that revolution included the slaughter of bourgeoisie who resisted.

In major German cities, *Freikorps* ("free core") militia gangs roamed free; often they were able to murder political opponents who frustrated their ambitions. Historians suggest that in the early 1920s, while Arendt finished high school and entered university, between 50,000 and 400,000 *Freikorps* troops organized throughout Germany, all of them itching to overthrow the civilian government.[26] Adolph Hitler's "Beer Hall Putsch," on November 8–9, 1923, which nearly succeeded in giving him control of the government of Munich, was a *Freikorps* campaign. His avowed aim was to consolidate his troops and march on Berlin, there to overthrow the elected government. An equivalent American story would be a massive army in Philadelphia, led by a renegade American general, prepared to march on Washington in 1933, to dissolve the Congress and depose President Roosevelt. Between the war's end in 1918 and the Beer Hall Putsch in 1923, Hannah Arendt came of age. During these years the cornerstone of Nazism was laid in Germany.[27]

As usual, when we try to understand a complex historical event we must go backward in time, seeking to discover the conditions that made it likely. To learn why Hitler nearly succeeded in what seems a bizarre attempt to overthrow the government, we need to understand Munich in 1923. To do that, we have to consider an event that preceded it by five years—the Revolution in Bavaria in 1918–1919, led by the radical Left against both the old regime and the moderate Socialists. After a series of crises and gun battles, reactionary and regime forces defeated the leftist groups that controlled Munich. The Revolution of 1918–1919 "led to the development of counter-forces and to the rise of a new generation of anti-Red leaders of whom the most immediately imposing was perhaps Colonel Franz Ritter von Epp, while the truly crucial figure in this wave was the still insignificant Adolph Hitler."[28] That Arendt and others found themselves yearning for stability and calm as they confronted terror does not seem surprising.

I prefer Lincoln to Heidegger. How might one justify this choice? To what evidence can I or an opponent who favors Heidegger appeal to argue our case? Lincoln is an American who writes in English; Heidegger is a German who writes in German. Lincoln is part of my boyhood, part of the American mythos, and typically cited as our greatest president. He is part of my idealized past. To go beyond that accidental feature of my story and to argue that Lincoln pertains to everyone, Iraqis as well as Americans, is to engage in a difficult and important task. It plunges us into longer and more inclusive conversations. That too, counts as an advance but it will not be *into* the depths, it will be *across* boundaries.

2. MAGNIFICATION AND CULTURAL OBJECTS

FANTASIES OF DEPTH:
MAGNIFYING CULTURAL OBJECTS

In the previous chapter, I argued against the notion of "depth" in depth psychology and similar forms of humanistic inquiry. It serves little purpose when we use it to think about progress in the humanities. I have not concluded, therefore, that there can be no progress in humanistic enterprises. On the contrary, we can point to numerous moments of humanistic success. They are not cumulative in the way that science is. This demarcates something important about humanistic enterprises, namely that they begin anew with each new generation.

Of course, we can draw on humanistic traditions. In subsequent chapters, we will consider Greek statues, theater, sculpture, and other forms of artistic expression as each developed over many hundreds of years. Sophisticated art historians can articulate these developmental stories. Using various criteria, such as the style of drapery in Greek statues, they can establish chronologies and show how later pieces derived from earlier pieces. They cannot then go on, however, and prove that the later are better than the earlier. Reflecting on the notion of progress in the sciences versus the arts, I noted that in the arts (a) latest is rarely best and (b) "best" always denotes comparison against similar expressions.

In chapter 1 I asked the reader to look at a slide show, "Magnification." These slides show seven examples of magnification. Some of these seven are natural objects; the others are cultural objects, the things that humanists study. Unlike natural objects, cultural objects will not yield to rigorous forms of magnification. We can use this difference to distinguish cultural objects from natural objects:

Natural objects can sustain magnification.
Natural objects do not convey meaning through horizontal linkages.
Natural objects extend vertically as well as horizontally.

In contrast to natural objects, the kinds of things that interest humanists—such as myths, self- and group narratives, art, and dream images—are cultural objects:

Cultural objects cannot sustain magnification.
Cultural objects convey meaning through horizontal linkages.
Cultural objects require contexts.

This exercise may appear trivial. Few art historians, for example, would argue that they could learn more about Michelangelo's genius by examining his marble with a microscope. Nor would most historians find it useful to get biochemical assays of the ink that Abraham Lincoln used to compose his great speeches. On the contrary, we esteem most those historians who can place Michelangelo and Lincoln in their historical contexts and, perhaps, compare those historical contexts against our context. Because our context differs from those of our parents, we will always need newly revised readings of Abraham Lincoln, just as Christians always need newly revised readings of the New Testament. These great tasks may require a lifetime of work. While we may use metaphors about digging into history, implying a vertical intensity, I suggest that the vast majority of humanistic study is horizontal. Attempting to carry out vertical analyses leads to gross errors. To illustrate this claim, I consider multiple instances of humanistic study—a series of cases from art history, studies of the effectiveness and limits of psychotherapy, comments on the novels of John Updike, and creation science.

HORIZONTAL ANALYSES IN ART CRITICISM

As a test of my slide show thesis, I examine a random set of essays in a recent *New York Times* art section. The first is a review of a show at the Philips

Collection in Washington, D.C. The curators mounted prints by the famous Japanese woodcut artist, Utagawa Hiroshige, and compared them against Western paintings done by artists whom Hiroshige inspired when his work was discovered in the mid-nineteen century.[1]

Comparing the European responses to the Japanese prints, Ken Johnson says, "It is easy to see how Modernists from Manet to Bonnard could find in the lucidity and technical and formal economy of those Japanese artists inspirational guides for escaping the suffocating conventions of Beaux Arts and Victorian painting."[2] This strikes me as an intelligent and plausible thing to say about Modernist painters like Paul Cézanne. For he and many others did find the conventions and rules of the academy too constricting. The academics' notion of progress and rules constricted Cézanne and others; they rebelled against these rules with profound self-consciousness. The clarity and lucidity of the Japanese print, as Johnson says, seems to have helped the Modernists free themselves from the past and make new art. We count that as progress; but, again, it is wholly unlike the linear progress we find in normal science. When Paul Gauguin removed himself from France and fled to Tahiti to produce there a new kind of primitivism, that too was an advance, but the direct opposite of linear progress.

Speaking of Gauguin, we recall one of his famous sayings: "La couleur qui est vibration de même que la musique" (Color like music is a vibration).[3] This yearning to find parallels between colors, feelings, and music animated Gauguin's reflections on his later work. It is a fascinating idea, subject of numerous subtle studies of the concept of *synaesthesia*. Looking to the top of the page of the same *New York Times* issue, just above the essay on Hiroshige, we find a piece on synaesthesia: "With Music for the Eyes and Colors for the Ear." Its author, Michael Kimmelman, compares various efforts to meld music, painting, and stage together.

Chief among these protagonists was Richard Wagner, who attempted to create what he called a *Gesamkuntswerk* ("total art"): "Like Wagner's Gesamkuntswerk, the dream of making one art that's like another is just a utopian fantasy, born of a peculiar modern impatience with art's limitations and a misplaced notion that, like science, art needs constantly to advance or else become irrelevant."[4]

By associative linkages to Richard Wagner and his concept of *total art*, we are drawn into yet another realm of complexity and context. For looking back at Wagner's writings about music, we recall also his virulent anti-Semitism; we cannot forget how closely he linked one with the other.

In his lengthy essay "The Art-Work of the Future" (1849),[5] Wagner forges a dense, highly speculative essay on Science, Art, and Truth—each word capitalized to designate its central and enduring importance. Drawing on his reading of Arthur Schopenhauer and his own views of German and English arts, Wagner announces that Art is of equal dignity to Science:

> Man will never be that which he can and should be, until his Life is a true mirror of Nature, a conscious following of the only real Necessity, the *inner natural necessity*, and is no longer held in subjugation to an *outer* artificial counterfeit,—which is thus no necessary, but an *arbitrary* power.

Science is that human endeavor that examines the realm of necessity; Art investigates an identical form of necessity, that which originates from within, "the inner natural necessity." False artists, such as Mannerists and Modernists, submit their work to the whims of fashion, taste, and superficial culture. Even those artists who work on behalf of the powerful in religious, state, and other centers of power obliterate the origins of art. That origin, Wagner says, is in the People, the Volk.

The Volk are not everyone who happens to live within a given geopolitical space. Nor is the Volk the totality of Christians, for example, or human beings. On the contrary, the Volk are

> those men who feel a common and collective Want [*gemeinschaftliche Noth*]. To it belong, then, all of those who recognise their individual want as a collective want, or find it based thereon; ergo, all those who can hope for the stilling of their want in nothing but the stilling of a common want, and therefore spend their whole life's strength upon the stilling of their thus acknowledged common want.

This wanting is a deep feeling; it arises from a common, collective need, a collective soul driven by genuine loss and satisfied only by an art that addresses those losses and needs. A total art, a *Gesamkuntswerk*, would be one that brought together the plastic arts, the arts of speech, poetry, drama, stagecraft, musical theater, opera, and architecture to create a singular experience. A *Gesamkuntswerk* would address that deep need; it would be a total answer.

It might prove instructive to dissect Wagner's essay more carefully and to compare it against the diagnosis of Germany's ills that Heidegger offered ninety years later. As it stands, I find it difficult to know what sentences like these mean:

> The real Man will therefore never be forthcoming, until true Human Nature, and not the arbitrary statutes of the State, shall model and ordain his Life; while real Art will never live, until its embodiments need be subject only to the laws of Nature, and not to the despotic whims of Mode. For as Man only then becomes free, when he gains the glad consciousness of his oneness with Nature; so does Art only then gain freedom, when she has no more to blush for her affinity with actual Life.

This seems to mean that, according to Wagner, there is a single Human Nature that requires a single, right relationship to Nature; somehow by 1850 or so, this right relationship has been harmed by mere fashion, versions of Modernism have distorted it. Wagner's task was to describe the deviations and correct them by creating a total artistic experience, something that he attempted to do when he constructed his theater-temple at Bayreuth in the 1870s.

It would be equally important to see how these exhortations pertain to Wagner's next essay, "Das Judenthum in der Musik," written two years later. In that essay, Wagner expounds upon the theme of "natural music" by commenting on what he says are inherent limitations to Jewish appearance, talent, taste, and therefore their inappropriateness for *Gesamkuntswerk*.[6] Questions about the origins, degree, and extent of Wagner's anti-Semitism pertain to German and American history, to the history of art, and to ethics, to name only three areas. For example, Wagner says that the Volk have an instinctive hatred of Jews: "For, with all our speaking and writing in favour of the Jews' emancipation, we always felt instinctively repelled by any actual, operative contact with them."[7] Why did he say this? This is a deep question in the sense of moral gravity; it requires careful, comparative study of Wagner's context, his other writings, the pervasiveness of European and Christian anti-Semitism, Wagner's relationship with Friedrich Nietzsche, and dozens of other tasks.

However, this is not a deep (vertical) question in the prosaic sense of my slide show. We cannot answer questions about Wagner's anti-Semitism by looking deeper into the inner workings of his music. We cannot subject the particular artifacts in which Wagner expressed himself to dozens of levels of magnification the way that we can examine slides of Australian coin or viruses. True, one could treat Wagner's music—when played—as a natural object. We could analyze its notes and sounds into smaller and smaller parts in the way that linguists can analyze phonemes into their phonetic components. However, very soon, this kind of analysis becomes reductionistic: the music disappears into lines on an oscilloscope, or numbers on a computer printout, just as the meaningfulness of phonemes disappears when we reduce them to their psychoacoustic parts.

When Wagner reduced his music to propaganda and revivalism, it ceased to be musical. For some, this amalgamation made it impossible to enjoy Wagner's artistry. Wagner complains that Jews merely listen "to the barest surface of our art, but not to its life-bestowing inner organism." One might counter this slur empirically by appealing to the facts; and one might counter it theoretically by noting that music is entirely on the surface: there is no inner organism.

Art criticism, a form of humanistic inquiry, seems to require vast erudition about numerous contexts and a willingness to ascribe motives and feelings to artists and to viewers. Thus, describing yet another comparative exhibition, this one of photographs by American artist Robert Mapplethorpe and Mannerist illustrations, Roberta Smith acknowledges sameness and difference: both sets of images revel in the stark form of well-muscled nude men. Yet, for all their similarities, Smith says, Mapplethorpe "never went near a crowd scene, usually photographing one person and rarely more than two in stark isolation. He inevitably emphasized them as three dimensional forms in empty, uneventful space. In the end it is hard to imagine him owning Mannerist prints."[8]

Even if the assiduous biographer discovered that Mapplethorpe indeed owned Mannerist prints, Smith's statement remains interesting. For it describes the critic's response to the comparison of the two forms of art. It might still be hard for Smith to imagine Mapplethorpe's sensibility enjoying the crowded scenes of the Mannerist prints. A great or influential critic may contradict the artist's self-understanding and retain his or her standing. The work of art, like numerous other cultural artifacts, is open to relocation, to reinterpretation and placement in another part of the cultural space. When Mapplethorpe or T. S. Eliot comment on their own work, they enter into the discourse of criticism and their pronouncements have no additional weight or authority. They must contest with other critics.

In a similar way, a brilliant humanist thinker might persuade us to align ourselves with neo-Kantians or neo-Freudians, for example. However, as we've seen, in the face of these apparent advances another thinker may rise up and refute Kant or Freud. It is surely important that a humanist may become great by interpreting Kant, Freud, or Plato when no such accolade befalls the ordinary scientist. One might become a great teacher of Darwinian theory and expert on Darwin's life, but these are not scientific achievements. Happily, some are both great scientists and great teachers; the latter designation places them among humanists.

PSYCHOTHERAPY: PART SCIENCE, PART HUMANITIES, MOSTLY ART

We know that psychotherapies work (as do many forms of religious cure). Yet, after numerous attempts, no one has identified the medical cause for success. On the contrary, psychotherapy is a mixed form of medical concepts and humanistic enterprise. If researchers are able to identify a medical cause for a specific malady, treatment for it migrates from psychotherapy to medical psychiatry. Having looked for and not discovered psychogenic factors that produce schizophrenia, for example, most contemporary psychotherapists categorize it as a biological disease.

Psychotherapy works, but it does so in general ways. This suggests that the "disease" for which it is the cure is also general. That is, some parts of psychopathology are context dependent, and other parts are not. If the disease is real but context dependent, then it must be shaped by both personal and cultural categories of self-understanding. The story that patients tell themselves and the story that cures them of their malady must conform to cultural norms.

Outside of psychoanalysis, many other schools of therapy have grown up. Adherents of each school have offered their own renditions of both what causes psychopathology and how to cure it. While they differ with regard to which factors they name as causes, most of these authorities agree that psychopathology is a kind of illness. Therefore, they adopt a medical model of psychotherapy: we first discover the causes of the disease, then, from that causal knowledge, fabricate a form of cure. If the medical model of psychopathology is valid, then eventually we should be able to locate these elusive causes and cure the pathology to which they gave rise. The attractiveness of the medical model is obvious: it builds upon the vastly successful model of modern medicine, which, since Abraham Flexner's report in 1910, has sought to base teaching and practice on scientific grounds. As Flexner put it, "Civilization consists in the legal registration of gains won by science and experience."[9] On these grounds, he oversaw a vast restructuring of medical education. He effectively outlawed hundreds of marginal schools that flourished before his report instigated reform.

From a new source comes an empirically grounded argument against this medical model and the search for causal factors that produce psychopathology.[10] In *The Great Psychotherapy Debate*, Bruce Wampold reviews hundreds of outcome studies of the effectiveness of various forms of "medical" psychotherapy, and thus the validity of their underlying theories of pathogenesis. Adherents of the medical model presume that, having identified causal agents, the psychotherapist uses precise therapeutic agents to counter them and to cure the patient's psychological disease. When Freud guessed that hysterical patients suffered mainly from their memories of forgotten traumata, that guess led him to emphasize recovering those memories by requiring his patients to dredge them up, and thus effect a cure. When contemporary adherents of cognitive behavioral therapy guess that many forms of depression derive from punitive self-talk, they focus their therapeutic efforts on confronting these internal dialogues, and thus effect cure. Medical therapists search for psychological causes and not spiritual ones; their master theory is one of specific cause and thus specific cures. In this way, they share a worldview with traditional shamans, or medicine doctors, who diagnose ethical lapses that caused their patients disease and that must be remedied to effect cure.

I use Wampold's work because it appears to be the most rigorous available and because it examines the effectiveness of competing schools of psychotherapy against one another. Regarding psychoanalysis, the bad news is that Wampold finds no evidence in favor of its specific claims to causal effectiveness. He finds no evidence that the specific factors that analysts cite as crucial to therapeutic work—such as insight into one's mental distress—are essential to positive out-comes. The good news is that he finds no evidence in favor of specific factors favored by other schools of psychotherapy. On the contrary, as I read him, Wampold names a paradox: many psychotherapies work well, yet their curative factors are general, not specific. In other words, psychotherapy is an effective form of treatment; it is not mere placebo. However, there is no evidence that favors the specific claims of a particular school of therapy. In the language of one commentator, this is the Dodo Bird outcome, citing a scene from *Alice in Wonderland* when the aforementioned bird announces, "All must have priz-es."[11] While amusing, this attitude overlooks that human beings can help other human beings with their serious emotional problems using a variety of tech-niques and following a variety of ideologies. That suggests something important about human adaptation.

In each of these schools, adherents claim to designate specific causal factors, what Wampold calls "specific ingredients," as the agents responsible for cure. Since psychotherapy, as an aggregate, has an effectiveness rating of nearly 80 percent, these are significant claims. On the one hand, evidence is strong that psychotherapy works well. On the other hand, there is little or no evidence for the medical model: "If the medical model provides a useful framework for con-ceptualizing psychotherapy, then evidence should suggest that the specific ingredients are responsible for the benefits of psychotherapy." Indeed, summing up his rigorous review of outcome studies, Wampold says that the combined evidence for specific ingredients hovers between zero and one percent: "Decades of psychotherapy research have failed to find a scintilla of evidence that any specific ingredient is necessary for therapeutic change."[12]

If we agree that there are no specific curative ingredients and that therapy works, then there must be general elements, what Wampold calls contextual elements, that distinguish effective psychotherapy from ineffective psychothera-py. Here Wampold joins Jerome Frank, who made this point some fifty years ago in *Persuasion and Healing* (1961).[13] These common factors are

1. an emotionally charged confiding relationship with a helping person
2. a healing setting that involves the client's expectation that the profes-sional will assist him or her
3. a rationale, conceptual scheme, or myth that provides a plausible, al-though not necessarily true explanation of the client's symptoms and how the client can overcome his or her demoralization
4. a ritual or procedure that requires the active participation of both client and therapist and is based on the rationale underlying the therapy

Not accidentally, Wampold uses the terms *ritual* and *myth*, which derive from religious studies and the anthropology of religion. Indeed, Claude

Lévi-Strauss, an acknowledged modern master, made this point in a famous essay in which he compares New York psychoanalysts to ancient shamans.[14]

Wampold's findings reinforce the line that I have pursued in a previous study: that psychopathology (as opposed to psychophysical diseases like schizophrenia) is not a medical entity.[15] Of course, this may be wrong. Next month someone may find the missing key, the single factor that has thus far escaped detection. If so, they will have brought psychotherapy into the orbit of the usual sciences, and psychotherapy would assume its place alongside the other scientifically grounded medical specialties.

In the meantime, assuming that Wampold's findings are valid, some interesting consequences follow. I summarize them by contrasting the four points above with the lived-world of the typical character-disordered patient, such as borderline patients.[16]

- An emotionally charged confiding relationship with a helping person— This is the hardest quality to create in a therapy with character disordered patients. Borderline patients, for example, spend their lives avoiding emotionally charged, confiding relationships with all persons. While the reasons they do so are multiple and complex, the consequence is simple: they refuse to share their most distressing emotions with another person. When they confide these feelings, especially their fears of abandonment and loss, therapy progresses. Indeed, the hardest part of the therapy is over. Getting to this point is not easy and many a therapy—regardless of school or technique—falters when the therapist cannot convince the patient to confide in him or her. Because persons diagnosed as borderline always manifest profound struggles with trusting others, to learn to trust the therapist counts as a major step forward. If the patient can trust one person, she or he can trust another, and then the world of friendship and its ability to heal opens up.
- A healing setting that involves the client's expectation that the professional will assist him or her—When the therapist can affirm that the borderline patient expects to be helped, the work proceeds nicely toward a successful outcome. "A healing setting" denotes the patient-therapist mix after much agonizing work has been done. Borderline patients have little or no experience of what "healing setting" means because it denotes an unknown world of joy. Thousands of painful interactions have taught them the prudence of their paranoid attitude. They expect, therefore, massive retaliation from everyone, especially those who promise them goodwill. Because these expectations of abandonment are typically validated by subsequent events, they deduce that this new relationship with the therapist will turn out like all the others. That a friend has been

loving and kind 199 times does not prove that he or she will be so on the 200th occasion; hence, the borderline person remains as paranoid about the 200th interaction as he or she was about the first. Even saintly friends and spouses will give up at some point, and this proves to the patient that his or her initial fears, repeated endlessly, were valid. Professional training of clinicians from all schools requires them to tolerate conflicts with borderline persons. Supervision, peer groups, one's own therapy, and other forms of counsel and learning take place, each designed to give beginning therapists support and to help them not run out of the room when the attacks begin. Theories of technique, the seemingly scientific part of training, have little effect on quelling the stormy feelings that borderline patients evoke.

- A rationale, conceptual scheme, or myth that provides a plausible although not necessarily true explanation of the client's symptoms and how the client can overcome his or her demoralization—This finding counters the wishes of a great many scientists and scholars convinced that the medical model is valid. It frustrates everyone equally since it denies that any of the major schools of therapy has designated a specific entity—the pathogenic belief, for example, or inhibited libido or irrational thoughts—as the causal agency that produces psychological suffering. On the positive side, this finding helps account for the interesting fact that all cultures at all times have created both conceptual schemes and methods to decrease psychic pain. These schemes are akin to religious myths, these methods are akin to religious rituals.

- A ritual or procedure that requires the active participation of both client and therapist and is based on the rationale underlying the therapy—In addition, these methods require that the interpreter plunge into a search for latent meanings. In the hands of a master clinician, these emerge sometimes and are plausible. When offered in the right context, these master metaphors and narratives lead to cure. Of course, these rituals and myths must be compatible with both persons' prior beliefs and value systems. If the dominant language and episteme is one of science, as in North America in the twentieth-first century, then that will be the context in which the curative myths and rituals must occur.

That patients agree with their therapists and share these master narratives, and that they get better, does not prove that the causal theory upon which these interpretations rest is valid. Wampold's results show us that there need not be any strong relationships between the master narrative and what science tells us. Religious cures, affirming vastly different meanings and offering vastly different interpretations, help change people's distress level. These cures do not validate the ontological claims made by religious experts, just as successful psychothera-pies do not validate the therapist's medical model of pathogenesis.

On the contrary, the task becomes one of ruling out competing interpreta-tions: clever readers and interpreters can find numerous connections between a patient's story and their own associations. If we agree with Freud that there are well-formed but hidden connections between the patient's story and the patient's early family life, we need ways to arbitrate between competing interpretations.

This takes us back to the job of validating hermeneutics. This job is not much fun. Having taken part in hundreds of case consultations, in perhaps one or two did the presenter focus upon validating therapeutic interpretations. It is more enjoyable to exercise one's imaginative capacity, to generate novel ideas, new readings, and insights, than it is to obey the killjoys who insist upon rigor. This also denotes the artistic side of interpretation and is one of the core values of psychoanalysis as investigation: that we eschew any answer that denies the reality of dual experiences. As literary critics have long known, deciding what will count as relevant text and nonrelevant text dictates the nature of one's dis-coveries. Unless the critic merely retypes the novel, for example, he or she must choose a way to find the work's essence, or at least, its essential features. For the surface of a work of art, the mere surface, one might say, is what demarcates great works from trivial works. The psychology of art that reduces it to deep themes and refuses to acknowledge the surface goes awry. Describing his work with a computer scientist, M. Gardner says:

> However much we analysts may differ regarding the depths—the distal events—we consider essential to explore, we differ still more, I believe, regarding the surface, the proximal events, from which we prefer to set forth on the way to the distal. It is to that choice of proximal events, a choice that strongly influences the directions of our subsequent soundings of the depths, and therefore of our fact findings, that I want now to invite attention. I want especially to invite attention to the ways in which the choice of where to begin reflects a never entirely successful effort to promote a relatively spontaneous unfolding of events rather than a forcing into the areas of our preference.[17]

Gardner's patient demanded that he clarify the logic and meaning of his feelings before they occurred. This demand made it impossible for the patient to free associate. Free association, Freud's earliest technical requirement, is one of those "relatively spontaneous unfolding" processes that may lead to novel insights. The faith of the psychoanalyst, expressed well by Gardner, is that decreasing the patient's control over the patient's mind is a good thing because it will make the spontaneous unfolding of events more likely. Hence, from Freud onward, analysts have encouraged free association. From most patients' perspectives, losing control over their minds and words is a bad thing. To lose control means to fall apart, to collapse into chaos, to be formless, to be covered in feces, or some other mix of shameful and disorganizing experiences. In the colorful language of French psychoanalysis, to not control the mind is to evoke the possibility of blitzkrieg affects.[18]

The value system that champions free association is similar to what I called "high art," in my discussion of aesthetics. "A relatively spontaneous unfolding of events" names the therapeutic goal of asking patient and therapist to tolerate the ambiguity of choosing one part of the surface or another as a starting point. In the paradoxical language of Freud's discovery, "free associations" are never free because unconsciousness needs and the shaping experience of transference mold the patient's thoughts. Earlier, archaic, and primal experiences that are driven by nonconscious perceptions and reactions trigger later, advanced, and learned defensive responses.

Much modern art begins with the abandonment of fully controlled expression. For example, consider a celebrated moment in European cinema, the funny three-way dance in Jean-Luc Godard's 1964 film *Band of Outsiders* (*Bande à Part*). Two somewhat goofy young men and a beautiful young woman plan a robbery; then a spontaneous dance occurs:

> What prevents this melancholy lark from becoming coy, a preview of shallow, post-modernist gamesmanship? First of all, there is Godard's inventiveness—what the film critic Robin Wood called the "sheer delight in creating that continually balances and colors our awareness of the near-despair that the film expresses and that makes us feel it, not as a litter of fragments, but as the expression of a single and strong creative impulse." The emotional tone—Godard's grave romanticism, if you will—draws every stray metal filing to its magnet.[19]

Art like Godard's that verges on the brink of chaos is not easy to enjoy. For like Gardner's analytic patient, Godard's viewers must give up their immediate needs to know what lies ahead of them. Pretty good art, like pretty good movies, is satisfying because it is mere fancy, that is, as Ruskin says, our wishes are fulfilled and we are not disturbed. One can see a war movie like *Pearl Harbor* (2001) and fall asleep easily that night. However, watching a great movie like *Grand Illusion* (1937) is disturbing; sleep may not come easily. Great or high art escapes our control and thus threatens us with genuine novelty. Contrary to our wishes, control slides over to the artist.

Commenting upon Godard's celebrated movie, critics employ brilliant metaphors to capture their feelings about the film: Robin Wood's metaphor of a "litter of fragments" is what one experiences on first confronting a Godard film. Extending Wood's metaphor, Philip Lopate adds another image of organizing power by way of the oxymoron "grave romanticism" and the nice conceit that this quality of Godard's imagination "draws every stray metal filing to its magnet." With sufficient artistic and intellectual force, nothing in the film need feel fragmented, out of place or stray.

While we can admire this in a movie director, it is less appealing in a psychotherapist. Godard's talent and mastery let him manipulate us, the viewer, into sharing his vision, his frame upon these three characters. Within the darkened theater, metaphors like "grave romanticism" help organize what could otherwise be incoherent and unappealing. However, outside the darkened theater, Godard's forcefulness is less appealing, as evident in his later films, which are overtly essays on Marxism as the solution to the crisis of art and politics.

In a similar way, Gardner can champion losing control of one's words because his analytic faith tells him that he and his patient can handle whatever emerges. He will be calm, at least calm enough that he remains in contact with his patient's experience. Gardner can say this only after he has lost contact with his patients' moods enough times to have learned how to stay on course. The expert therapist feels relative calm amidst these storms. When the patient perceives these manifest truths radiating from the therapist's voice and demeanor, therapy can continue and prove helpful. Perceiving these "superficial" signals is inherently soothing to patients; it grounds their hope that fragmentation and dissociation will yield, eventually, to order and meaning. Like good psychotherapy, good art pertains to the surface of things also.

JOHN UPDIKE, RABBIT RERUNS

Clinicians of all types must be optimistic: we believe that we can name what is not yet named. At this level of human interaction and cure, the medicine expert must grasp the intimate social-cultural context in which mental illness and distress make themselves visible. In a similar way, it takes persons of genius to articulate a certain kind of emotion that arises in particular circumstances constrained and defined by particular contexts. Thus, for me, John Updike captures feelings rooted in specific, American, middle-class lives of the late 1960s through the late 1990s. He chronicles them in his four volumes on the life of "Rabbit," the nickname of Harry Angstrom. Harry earned that name in boyhood when he was a swift runner and when he hoped to escape his small town life and, vaguely, his fate. Because his fate also involves his death, Harry's ruminations constantly come back to that theme.

We met Harry in *Rabbit, Run* (1960), have followed him through middle age in *Rabbit Redux* (1971) and *Rabbit Is Rich* (1981), and finally into old age in *Rabbit at Rest* (1990). In the latter story, Rabbit struggles to tolerate the awful truth: he's no longer fast, he has become fat, he has lost his sexual vigor, and he is a grandfather with a bad heart. Readying himself for a day with his grandchildren, Harry watches his wife dress, then he dons his old-man ensemble: "He looks at himself in the mirror that Janice's image has vacated and is stunned, deep inside, by the bulk of what he sees—face swollen to a kind of moon, with his little sunburned nose and icy eyes and nibbly small mouth bunched in the center, above the jowls, boneless jowls that come up and put a pad of fat even in front of his ears" (*Rabbit at Rest*, 74).

Like an anthropologist's deep descriptions of a foreign culture, Updike portrays with exactness a closely examined life. Rabbit stares into the mirror looking for his lost youth, trying to find the vigorous man who once had a strong jaw and lithe body. In those days he had a sense that time (and sexual healing) lay stretched out, if not endlessly, then at least farther than he could comprehend. The pleasures that Updike affords us are analogous to the pleasure and curative effect of rituals: a master of our culture has given us a pattern language for recognizing and locating ourselves in an otherwise empty universe. Facing certain death is painful and disorganized, if like Harry, we are left to ourselves. We can empathize with Harry but we don't share his despair. We want to go on reading the book because it gives us intense pleasures not available through any other means. As Joyce Carol Oates put it, "*Rabbit at Rest* is certainly the most brooding, the most demanding, the most concentrated of John Updike's longer novels. Its courageous theme—the blossoming and fruition of the seed of death we all carry inside us—is struck in the first sentence."[20]

By using her metaphor that death is a seed that blossoms into something flowerlike, Oates extends the power of Updike's art to reshape our fear of death. This would not occur if we did not love Harry, and we would not love Harry if Updike had not taken us through his anguish, lust, hope, and a hundred other feelings named for us in the course of this masterwork. Woven from thousands of exact representations of lived experience, *Rabbit at Rest* lets me and those who share my world think about the seed of death planted within us from the safety of aesthetic distance.

Near the novel's end, Harry has been pushing himself to play one-on-one basketball against a young man. He collapses with a massive heart attack onto the schoolyard: "Seen from above, his limbs splayed and bent, Harry is as alone on the court as the Sun in the sky, in its arena of clouds" (421). We join the novelist (and God) in looking down from the heavens toward earth and contemplate there the crumpled shape of a man, alone in the world. Harry is also like the Sun, another kind of round ball, solitary in another kind of arena. We also remember that the first novel, *Rabbit, Run*, opens with Rabbit watching some boys play a game of pickup basketball. In that earlier scene, Rabbit feels the loss that he is no longer the star athlete: "He stands there thinking. The kids keep coming, they keep crowding you up" (*Rabbit, Run*, 9).

Good rituals, like good novels and good symphonies, announce themes, develop them, work out variations, and then resolve them for the benefit of the participants. These need not be grand. A delicious moment occurs in *Rabbit, Run* when Harry reflects upon a homily offered on the television show *The Mouseketeer Club*. Jimmy, the chief Mouseketeer, cites the ancient Greek proverb "Know Thyself"—"Now what does this mean, boys and girls? It means, be what you are" (14). No doubt to an anthropologist from Sri Lanka or another culture not permeated by the Disney company, this whole "mouse" business, in which prepubescent children wear mouse ears and sing and dance, would appear mysterious. What links mice to American children to the Greek phrase *Gnōthi seauton* inscribed on the temple at Delphi?

For those shaped by the Mickey Mouse Club this is an outlandish question. We *know* that we should listen; we *know* that Jimmy, the village elder, is telling us something for our own good as future members of the American economy. The Greek call to self-knowledge, which Ernest Jones said Freud was the first to address fully, becomes, on the *Mouseketeer Club* a call for hard work and career planning: "God wants some of us to become scientists, some of us to become artists, some of us to become firemen and doctors and trapeze artists. And He gives to each of us the special talents to become these things, provided that we work to develop them. We must work, boys and girls. So: Know Thyself" (15).

How should we settle the dispute between Ernest Jones and Jimmy as to the meaning of *Gnōthi seauton*? Some would wish to talk with experts on Greek history and philosophy before they agreed with the writers on the Mickey Mouse Club. Yet, subtler than an appeal to scholarly authority is Updike's portrait of Harry and his wife, who are entranced by the chief Mousketeer's homily. Using terms of this chapter, Updike's empathic account is a form of horizontal research: he helps us locate the Greek phrase in an American ideology about work and its inherent nobility. Updike does not disdain his characters or Jimmy the Mouseketeer. On the contrary, he uses his art to illuminate them and in doing that reveals his deep affection for them. Like the Dutch painter Johannes Vermeer, Updike brings tremendous care to these domestic portraits. In one sense, they are nothing more than shallow bits of paint—or a few words—on a flat page. Yet, in a more important sense it is the surface of things that most matter. Our feelings and sense of self reside in specific smells, sights, sounds, textures, in all the things upon which we map our emotional lives.

Another form of search into the meaning of cultural artifacts is vertical: to examine the thing itself for its hidden meanings just as the anonymous web author searched the map of Washington, D.C., for clues to its Luciferic meaning. This form of search appears typically in occult efforts to read a text at deep levels. At these extremes appears the common wish to treat a sacred text as if it yielded an infinite amount of information. We see this wish enacted in the spate of books on the "Bible code" and in the recent upsurge in fundamentalist treatment of sacred texts.

An example of the first are best-selling books that purport to show that biblical texts predict the future. In one version of this story, through proper reading of the first five books of the Hebrew Bible, one can find predictions to key events in human history. According to the book jacket, "As God dictated the first five books of the Old Testament, He enclosed prophecies in a skip code—that is, every fifth letter in a sentence forms a word. The trouble is, the Code is so divinely complex, you need a computer to find it. Now that we have those, and author Michael Drosnin, you too can read God's secret messages in *The Bible Code*."[21]

We must reject this on logical as well as empirical grounds. On logical grounds, it is difficult to make sense of the proposition "predicted the future" since the future does not exist per se at that time, unless, that is, one imagines that everything that could occur is somehow referenced in these texts. If one asserts this, then the future is shaped entirely before it occurs and if that's true, then there is no choice and no human freedom. If the Bible foretells the assassination of John F. Kennedy, and we read this before November 22, 1963, it does us no good to alert the White House since the Bible cannot be wrong. If we did warn JFK and he avoided Dallas on that day, then the prophecy was in error, or we will need to reinterpret the text. More so, if the future is predicted and thus unalterable, it seems odd to lay down a set of rules of conduct, the very center of Torah, the Law. If our destiny is foretold, why struggle to obey the Commandments? If our actions, including obedience and disobedience, are already proscribed by God, where does human choice occur? In addition, why could God curse Adam, Eve, and the Serpent for disobeying him (Genesis 3:14–19) when they had no real choice?

To address these obvious problems, one must make special cases, each of them more tenuous and unlikely than the first. For example, one might argue that the Bible code only specifies the names of a few great people and the assassinations of only a few, relevant leaders. "Great" and "relevant" are established post hoc and then found in the text.

The empirical grounds for rejecting Bible codes and its numerous cousins are substantial. The initial claims for the validity of the Bible code were based on vague demonstrations in which advocates manipulated the Hebrew letters of the Genesis text. Then, armed with the names and dates of famous rabbis, they sought to find those names referenced within the array of possible combinations. Like similar accounts of extrasensory perception, Bible code advocates claimed that they found many more matches than one would expect if there were no code. Thus, apart from the shaky premises and ad hoc nature of these efforts, everything depends upon these discoveries being unlikely from a statistical point of view. Serious and respectful refutations of the Bible code have shown that these attempted linkages fail and that the experiments that produced the first findings cannot be duplicated.

In addition, we must reject these claims on lines of information theory. If the Bible code technique were valid, that would mean that Genesis, a finite text of some 78,064 Hebrew characters, specifies the actions of billions of persons over thousands of years. If we agree that at least eight billion people have lived since 700 B.C.E. and each person lived on average twenty years, then Genesis encodes the actions and destinies of 160 billion person-years (20 years x 8,000,000,000 people).[22] Thus, each single character in Genesis designates some 2.05 million person-years (by name and by dates). This seems unlikely. Expanding the Bible code to the remaining four books of Torah doesn't change the odds very much. To defend the Bible code as remotely plausible (and ignoring the problem of predicting future events that extend through trillions of possible permutations), religious authorities must designate a much shorter list of persons who really count—such as famous rabbis or, if Christian, famous and infamous popes. (In one experimental trial, names of the rabbis could include up to eleven variations; hence, if one couldn't find the rabbi under his usual name, one had ten more chances.) With flexible and arcane rules, one can always find a way to impose upon a semistructured array of data one's favorite pattern. (This occurs typically when a group of people examines slide E in the slide show; or when they try to find meaning in the slides of cultural objects seen at very high levels of magnification. In both cases, there is too little actual information to support inference. If we must find a meaning, the only source is projected fantasy.)

On empirical and logical grounds, no known mechanism could account for the ability of the authors of the text to know anything about events that lay ahead some two thousand years. If one claims that God is the author, of course, then the equation is changed and miracles abound. That still does not explain how a finite text could designate many billions of discrete facts. This too might be a miracle, which by definition is contrary to natural law. Of course if one asserts that God wrote the text and that God makes it do miraculous things, then anything goes, and with it human understanding and any possibility of science.

It is much more likely that here, as in other instances, we find religious motives. (Many of the original scientists who propounded the Bible code were orthodox Jews.) *The Bible Code* is a form of numerology, not science. Like other forms of occult practice, it illustrates a profound wish to treat a cultural artifact as if it could sustain an infinite series of reductions, examinations, and rereadings—that is, magnification without end.

This wish and this error reappear sometimes in the celebration of introspection as capable of yielding an unending series of new insights. For example, in 1909 William James defended psychical research: "The only form of thing that we directly encounter, the only experience that we concretely have, is our own personal life. The only complete category of our thinking, our professors of philosophy tell us, is the category of personality, every other category being one of the abstract elements of that."[23] By emphasizing the claim that we know best our personal experience, James lends support to a larger claim that we can investigate this "complete" category by introspecting its contents.

Robert G. Jahn and Brenda J. Dunne cite the same passage with approval[24] and add this comment from Henri Bergson: "Science and metaphysics therefore come together in intuition. A truly intuitive philosophy would realize the much-desired union of science and metaphysics. While it would make of metaphysics a positive science—that is, a progressive and indefinitely perfectible one—it would at the same time lead the positive sciences, properly so called, to become conscious of their true scope, often far greater than they imagine. It would put more science into metaphysics, and more metaphysics into science."[25] If this were true, one could plumb the depths of individual experience and find new truths without end. As Bergson notes, this would make metaphysics into a progressive science. This is precisely what cannot happen using any introspective technique.

If they search hard enough, clever persons can impose an interpretative theme upon any array of minimal data. Like slide E—which is of nothingness—the hypermagnified photos of Lincoln and Susan B. Anthony have so little actual structure, so little information, they are like clouds in the sky, subject to infinite "readings." Numerologists and other psychic professionals have made a living doing this for millennia. The danger of the imposed (projected) solution has worried scientifically minded psychotherapists from the beginning. In his zeal to solve the problem of the meaning of dreams, did Freud project his theory into his patients' materials, thus finding there what he had unconsciously planted? Paul Meehl, a distinguished methodologist, called this the "Achensee Question."[26] Following a therapy hour in which Meehl made an insightful interpretation to a young man, the patient dreamt about a *Time* magazine story on the Burmese Prime Minister, "U Nu."[27] In elaborate detail, Meehl shows that this odd dream element meant "You knew," that is, the patient's dream denoted his shameful feelings that Meehl knew something about him that the patient did not. Meehl persuades an unbiased reader that here we find an unconscious complex of ideas revolving around themes of aggression, castration, competition, and transference. In this sense, Meehl satisfies a reasonable criterion of plausibility.

However, as Meehl notes, even in clear instances like this it is not easy to formulate the common denominator or the red thread that ties his patient's thoughts together: "We have to discern what is common in the blocks of verbal output, but 'what is common' resists any simplistic semantic or syntactic categorization."[28] To rectify this situation Meehl proposes a general research strategy, "Topic Block Theme Tracing," which asks expert analysts to find the thread of a given set of verbal outputs from specific patients: "When these batches of rated themes are colligated in a single table, one now reads horizontally instead of vertically, to see whether the thematic 'red thread' is apparent."[29]

When Meehl showed that his patient had used the *Time* magazine cover story about U Nu to signify his distress in the previous hour, this seems an irrefutable instance of Freudian insight. However, it proves difficult to specify the conditions under which one could repeat this clever moment. Our task, again, is to find a method that bridges the gap between brilliant but idiosyncratic hermeneutics and generalizable and replicable methods. When this gap is large, psychotherapy remains more art than science.

Sometimes an unconscious idea is singular and, when discovered, admits of no other possible explanation. Meehl, a consummate methodologist, shows that in this incident he correctly interpreted his patient's seething anger at him because Meehl seemed to favor another patient. Meehl's interpretation of his patient's thoughts about "You knew" and "U Nu" are as persuasive as Freud's brilliant deductions about the Rat Man's behaviors (1909).[30] At other times, while we are convinced that there is a pattern inherent to the patient's behaviors, verbal and otherwise, we cannot specify it. After listening to someone for, say, two hundred hours, we recognize this pattern, but cannot always label it or isolate a single theme that gains the status of solution to the puzzle in the way that idealized detectives solve their cases once and for all.

Given this point of view, the faith of the analyst is that it's worth trying to find the connection between a dream and the day's events that preceded it. We cannot believe that dreams are nothing but the machinery running down, nor can we believe that dreams foretell the future. (Though one does find hints in Freud and a few others that this version of dreams as sources of occult knowledge appeals to them.)[31] Thus, most analysts and most patients are located somewhere on the spectrum of beliefs and practices that recognize that dreams are important and that they can be interpreted, even if not in definitive ways. We cannot accept the advice of the extreme skeptic that dreams mean nothing, nor can we affirm the wishes of the extreme advocate that dreams mean everything. We have to locate ourselves somewhere on that spectrum. Establishing just where the line is between plausible and implausible, however, is not an easy thing to do and will not satisfy skeptics on either side.

SCIENCE, ART, METAPSYCHOLOGY, AND MAGNIFICATION

Metapsychology is the attempt to give a scientific grounding to human experience. It centers upon terms and concepts taken from various discourses; some are similar to psychology (such as literary criticism), others are far removed (such as electronic computers and neural networks). Metapsychological discourses range from neurosciences, to genetics, to sociobiology, to metaphysical claims about spiritual essences or stories about past lives and such. Both a well-funded research scientist studying brain processes and a lonely spiritualist thinker telling stories about the gods practice metapsychololgy.

Christian authors who cite the Holy Spirit as a causal influence in human lives are engaged in metapsychololgy. Advocates of anti-Darwinism who use the Bible as a counterpoint to science also advance a metapsychology, one built upon biblical inerrancy and the alleged perfection of Scripture. Thus, authorities of the Institute for Creation Research (ICR) state that their purpose is "to discover and transmit the truth about the universe by scientific research and study, to correlate and apply such scientific data within the supplemental integrating framework of Biblical creationism, and to implement them effectively in traditional graduate degree programs with standard core curricula in science and education."[32] Here, a fundamentalist reading of the Hebrew and Christian narratives serves as an "integrating framework." This means that ICR members who are in good standing must align their use of scientific claims to the unchanging truths of the Bible, as interpreted by the proper authorities.

It is easy to dismiss the Institute for Creation Research as aberrant, a small player compared to the institutions funded by immense public and private support. The latter institutions favor the modern synthesis of Darwinism and genetic theory. While there is little danger that the ICR will siphon off funds from scientific medicine, or any form of science for that matter, the persistence of this form of thought is interesting. It has distinguished company.

When William James joined dozens of famous scientists and scholars in the American Society for Psychical Research, he advanced a spiritualist metapsychololgy.[33] James articulated a goal for psychology that it cannot reach, namely, that through intense examination of one's personal experience, one could advance knowledge. Using the shorthand of the slides, this amounts to saying that we can examine inner experience in finer and finer detail, discovering new information at every level. By maintaining this "magnification error," as I have called it, James can defend his religious beliefs against the encroachment of scientific discoveries about the mind. This faith in the infinite depth of personality emerges when James describes his vision of how future scientists would look back upon the reign of materialism dominant in the nineteenth century:

> The only form of thing that we directly encounter, the only experience that we concretely have, is our own personal life. The only complete category of our thinking, our professors of philosophy tell us, is the category of personality, every other category being one of the abstract elements of that. And this systematic denial on science's part of personality as a condition of events, this rigorous belief that in its own essential and innermost nature our world is a strictly impersonal world, may, conceivably, as the whirligig of time goes round, prove to be the very defect that our descendants will be most surprised at in our own boasted science, the omission that to their eyes will most tend to make it look perspectiveless and short.[34]

I have cast my lot with the scientists because I locate myself as a psychoanalyst in an American university in the beginning of the twentieth-first century. Like most of my colleagues, I am struck that for some five hundred years now, one form of general inquiry, what is known as the scientific method, has produced advances in understanding not produced by any other method. From the beginning, psychoanalysts have aligned themselves with science. Sometimes Freud had the temerity to claim that his discoveries and his methods counted as scientific; at other times he saw his discipline as headed toward an eventual marriage with scientific disciplines. Assigning Freud and psychoanalysis to the halls of science, or to the status of persons peering in from the outside, is less important than aligning psychoanalysis with scientific values. For the latter only require us to respect what the consensus of contemporary scientists count as fact and as demonstrated. Of course, the consensus may be mistaken; next year the consensus may change and everything said now will be jettisoned. That, however, would also count as one form of progress.

3. BACK TO FREUD, BACK TO THE GREEKS!

WHAT COUNTS AS PROGRESS
IN THE HUMANITIES?

It is easy to say what counts as progress in some endeavors: an economy progresses when it produces more wealth and better jobs, when the standard of living goes up and child mortality goes down. A science progresses when its theories explain more, with fewer anomalies, and it enhances our ability to control phenomena of interest. Medical science progresses when it helps increase the quality of life and eliminates a disease like polio. In all these cases we feel justified in using the word *progress* because we can cite agreed-upon measures of success. Poverty, ignorance of natural laws, diseases and scourges are bad; diminishing them counts as good.

It is much more difficult to say what counts as progress in those parts of life assigned to the humanities. In these areas, such as art, religion, education, and philosophy, what counts as success is not so clear. For example, looking back over the history of European painting, experts may agree that Impressionism differs from Modernism (which succeeded it) and both differ from Postmodernism (the most recent), but counting the last as progress is not affirmed universally. In his provocative studies of the effect of buildings on human happiness, Christopher Alexander, an American architect, argues that many of the buildings constructed in the last three hundred years are hostile, aesthetically wrong, and harmful to the human psyche.[1] Alexander may be wrong, and he has his share of critics. However, it's difficult to imagine a similar critique with similar authority of any discipline accorded the title "science."

The difference between the humanities and the sciences appears in the way each subject is taught. In the natural sciences instructors use current textbooks that they assume contain the best of previous scientific efforts. Because the sciences advance, more or less in a linear way, the latest is usually best. This is never said about humanistic disciplines like philosophy or literary criticism or some schools of psychology. B. F. Skinner, the most famous American psychologist of the late twentieth century, lamented this difference in *Beyond Freedom and Dignity* (1971). There he complained that humanists and even psychologists continued to read Plato and other ancients "as if they threw light on human behavior."

Humanistic disciplines are difficult to teach—especially in the glare of radical critiques like Skinner's—and even more difficult to advance. One reason for this is the lack of a consensus about what constitutes progress within the disciplines themselves. One finds numerous, competing, and mutually exclusive schools of thought. Prestige and authority reside sometimes with one school, sometimes with another. The same is true of borderline disciplines like psychoanalysis. Roy Schafer, a noted psychoanalyst, points out that we find no "Institute for Darwinism" alongside the numerous institutes for psychoanalysis. Darwin's work has been incorporated into the mainstream of scientific thought; Freud's has not (although it has had immense influence in many disciplines). An eminent psychoanalyst like Jacques Lacan can become famous by declaring that we must go back to Freud.[2] Martin Heidegger, one of the central figures of twentieth-century European thought, can say without embarrassment "Back to the Greeks," meaning, not back to Plato and Aristotle, but back those who *preceded* them, such as the pre-Socratic philosophers Heraclitus and Parmenides.

BACK TO FREUD, BACK TO THE GREEKS!

Heidegger illustrates the problem of progress in the humanities. For at the center of Heidegger's thought is the rejection of the idea of progress, especially in its twentieth-century guise in which technology and the march of science dominate consciousness. In Heidegger's reading of the history of Western philosophy, the beginning of thought was in the archaic Greek period when Parmenides, among others, asked not "what is knowable?" per se but "how goes it with Being?" While he locates the rise of science and rationality in the thought of Plato and Aristotle, Heidegger does not consider that a singular moment of progress. On the contrary, with the advent of normative reasoning came the loss of the original question: what is Being as a whole? Like Jacques Lacan, who lambasted post-Freudians for forgetting what Freud's courage revealed, Heidegger lambastes mere technicians who have forgotten the originating questions of Being. Not accidentally, I note, these deeply emotional metaphors—of rescue and retrieval—are prophetic and religious. Like numerous instances of reformation, recall, and reparation in European and American religion, this call is a summons back to an original unity, to an original Good. And like religious visionaries, Lacan and Heidegger produced passionate followers.

Carl Jung, Freud's most famous student, took up the same banner when he rejected explicitly the dominance of science and its ideal of progress. In his voluminous writings on psychology and religion, Jung says that something went wrong around the sixteenth century when European thinkers lost touch with the truths of prescientific eras: "From Descartes and Malebranche onward, the metaphysical value of the 'idea' or archetype steadily deteriorated. It became 'thought,' an internal condition of cognition."[3] Jung says that archetypes are archaic forms of thought, the very conditions for later concepts. Jung aligns his concept of *archetype* with Plato's notion of the "Forms" or *eidōlon* and distinguishes them from conscious, distinctive thoughts subject to logical and scientific examination.[4] Archetypes are forms of the instincts, Jung says, the foundational grammar of human self-understanding. Given their primordial nature, archetypes cannot be reasoned away by scientists and their adoration of progress. Because Western thinkers, even Freud, scorned the archetypes and adopted Descartes's methods and epistemology, Jung says that we have lost sight of their ubiquity and power.

Given these rejections of the ideal of progress, it is not surprising that those who associate themselves with Lacan, Heidegger, and Jung reject scholarship that depends upon ordinary notions of reasoning and proof. For the latter reveal the dominance of mere reason. They are symptoms of the very disease that each man sought to diagnose and cure. Like Friedrich Nietzsche, who preceded them and upon whom each man dwelt with intense devotion, Lacan, Heidegger, and Jung (and Freud) inspired similar devotion among their followers.

Ernest Jones, Freud's official biographer, could not resist linking Freud to the Greeks. Indeed, in Jones's estimation, Freud was the first person to obey fully the deepest Greek injunction, "Know thyself." In this way, Jones tells us, Freud completed the great circuit of investigation and thinking begun by the Greeks (whom Freud, like other Europeans, idealized). Jones says Freud's task was Herculean: "Few, if any, have been able to go as far as he did on the path of self-knowledge and self-mastery."[5]

Jones wrote this tribute shortly after Freud's death. We understand that grief may have pushed him to exaggerate Freud's stature. Yet, it is difficult to believe that Freud's work was greater than Newton's—or Einstein's. Less understandable is the degree to which psychoanalysts (among whom I count myself) and others have continued to lionize Freud some fifty years after his death. This rudimentary fact suggests that Freud means something for psychoanalysts that Darwin does not mean for biologists. How could Jones elevate Freud over every author and scientist since the Greeks? As we will soon see, Martin Heidegger made a similar claim for himself. Unless we reject Jones and Heidegger as kooks, it makes no sense to denigrate them. Rather, we should investigate their yearning to connect modern thought with the "beginnings" of thought in the archaic Greeks.

This commonality of yearning is part of the puzzle why humanistic enterprises do not progress in the way typical of the sciences. The pull toward idealization, the fascination with the personality of great figures is, I will suggest, another clue that the objects of humanistic inquiry are not structured like the objects of scientific inquiry. For humanistic pursuit includes the wish to find an answer to life in the character and thought of a great person, an exemplar human being who can show us how to complete ourselves. Not accidentally, this is precisely the tone of Freud's most influential book, *The Interpretation of Dreams* (1900), and Jung's most influential book, his spiritual autobiography, *Memories, Dreams, Reflections* (1961). Freud offered new ways to integrate parts of the self with other, hidden parts; Jung offers a way back to spiritual truths hidden by the rise of modern science and rationality.

Similar idealization occurs when Eugen Fink, a colleague of Heidegger's, describes his feelings about a course they taught together on the Greek philosopher Heraclitus. At the end of the yearlong seminar conducted in 1966, Fink summarizes some of his feelings about having had Professor Heidegger present: "The work of thought can be like a towering mountain range in stark outline, like 'the safely built Alps.' But we have here experienced something of the flowing magma which, as a subterranean force, raises up the mountains of thought."[6] This kind of adulation typifies much of the seminar's reflections. It echoes Heidegger's insistence that the earliest Greeks apprehended something true about Being and that this apprehension has the power to restore human beings to their proper place with nature and God. Deeply fond of short citations and sentence fragments—like his model, Parmenides—Heidegger tends toward prophetic utterances that sometimes meld into poeticism.

For example, in "The Fieldpath," a meditation on nature, Heidegger says: "The Simple preserves the riddle of the abiding and the great. Spontaneously it takes abode in men, yet needs a long time for growth. In the unpretentiousness of the Ever-Same it conceals its blessing. The expanse of all grown things which dwell around the Fieldpath bestows the world. It is only in the unspoken of their language that, as the old master of letter and life, Eckhart, says, God is God."[7] One may read this as a deep truth about the eternality of being that, when comprehended, gives modern persons a way back to the beginning. Or, one may read it as a tautology that is both universally and trivially true.

Given the density of these sentences and the poetic force of his language, it is not surprising that one can read Heidegger in ways that yield different pictures of his thought and of his character. The latter appears in a much-discussed speech. In January 1933, while the world witnessed the installation of Adolph Hitler as German Chancellor, Heidegger had just finished delivering two years of lectures on Aristotle and the pre-Socratics, particularly Anaximander and Parmenides. In April 1933, under the aegis of the new Nazi government, Heidegger became rector (or president) of the University of Freiberg. In May, one month later, he delivered his rectorship address, "The Self-Assertion of the German University" (*Die Selbstbehauptung der deutschen Universität*).[8]

As rector, Heidegger identified himself fully with the Nazi ethos and took upon himself all the trappings of a Nazi official. Numerous people have written about this episode, including his eminent colleague Karl Jaspers. Jaspers, a world-renowned philosopher and psychiatrist, had married a Jewish woman and remained, therefore, subject to Nazi race laws throughout the duration of the Third Reich. The extent of Heidegger's devotion to Nazism, his affirmation of "race science," his treatment of his great teacher, the Jewish philosopher Edmund Husserl, and the depth of his anti-Semitism are subject to heated dispute.

Separate from these disputes, we can say that Heidegger remained true to himself and the ideas already set forth in his magnum opus, *Being and Time* (1927). For, from Heidegger's perspective, German politicians were merely coming to insights about time and history that he had already articulated. While Heidegger was a Nazi, he was not sociopathic—at least the evidence does not yield a consistent picture of sociopathy. No doubt Heidegger saw the opportunity to become *Reich*-Rector as a happy one, but he did not alter his thought or conceptions upon achieving that post. (This may make one admire him the more or despise him additionally.)

On the contrary, we can read Heidegger's address as but another instance of his idealization of the archaic Greeks, interpreted, of course, using his categories. For while the civilized world watched with dread as Hitler ascended to power and democracy collapsed in Germany, Heidegger did not dwell on those epochal events. Nor does he consider directly recent German history, including Germany's defeat in the Great War and the rise and fall of the Republic.

In some four thousand words addressed to students who had suffered through the collapse of the Weimar government, Heidegger calls them back to their spiritual paternity. The students' place in the nation requires them to understand their mission, a mission given to them by their forbearers. These great ancestors are not Kant, Fichte, and Hegel—or Paul, Augustine, and Martin Luther. They are the archaic Greeks:

> Only when we submit to the power of the *beginning* of our spiritual-historical existence. This beginning is the beginning [*Aufbruch*] of Greek philosophy. That is when, from the culture of one Volk and by the power of that Volk's language, Western man rises up for the first time against *the totality of what is* and questions it and comprehends it as the being that it is. All science is philosophy, whether it knows it and wills it or not. All science remains bound to that beginning of philosophy and draws from it the strength of its essence, assuming that it still remains at all equal to this beginning.[9]

If this is true, then all science and knowledge depend upon philosophy, and philosophy began when the archaic Greeks reflected upon Being (*Ontos*). Thus, all parts of the university must orient themselves along this essential truth. Heidegger exhorts his students "Back to the Greeks!":

> Here we want to recover for *our* existence two distinguishing characteristics of the original Greek essence of science. Among the Greeks there circulated an old report that Prometheus had been the first philosopher. It is this Prometheus into whose mouth Aeschylus puts an adage that expresses the essence of knowledge: *techne d'anangkes asthenestera makro* ("But knowledge is far less powerful than necessity").[10]

But we are not called back to any old Greeks. For example, we are not called back to Plato and Aristotle, who wrote in depth about science and the problems of scientific progress. Rather, we are called to those Greeks who preceded them, the pre-Socratics. They were earliest, and earliest is best: "Only when we submit to the power of the *beginning* of our spiritual-historical existence" can we regain what is lost, the spiritual power of these earliest thinkers. This spiritual power resides in their character, not in refined thoughts and categories. Indeed, because we have no complete treatises from the pre-Socratics, we have no way to know what they were talking about. We have even less opportunity to know how they argued for their various claims.

Heidegger's call resembles numerous instances in the history of religion: a reformer rises up and declares that the current generation has failed to hear the authentic wisdom of the idealized past. This prophetic dimension echoes in Heidegger's celebration of the earliest thinkers and their utterances. This makes sense only if we believe that the power he ascribes to them was greatest in the first moments.[11] Like pieces of the True Cross, these earliest voices are nearest to the original truth or revelation and thus they are in closest contact with the original power. Heidegger wishes to link his philosophical discoveries to their thought; his new philosophy is the refinding, the uncovering of ancient truths. The two millennia of effort spent toward philosophical clarification since the ancient Greeks has not, according to the prophet, advanced self-understanding. The rise of modern thought and the modern sciences are not instances of unquestionable progress; on the contrary, they are part of the covering up, of forgetfulness of the confrontation with Being.

Heidegger calls his students back to the archaic Greeks because, he says, the Greeks had the courage to ask the deepest questions and grasp the deepest truths. Like them, Germans of all types and in all the callings will need to call upon similar courage. The Greeks, especially those whom Heidegger read in fragments, had a heroic attitude that let them avoid the turn to mere methods and technologies. As technologies, science and progress developed within a narrow horizon of function and comfort. These are not heroic values; they will not motivate German students to assume their heroic role in the new German state. Rather than celebrate the progress made by German chemists—and German-speaking physicists such as Albert Einstein, Max Planck, and Werner Heisenberg—Heidegger disparages those achievements as afterthoughts, the tinkering of men who have forgotten the original power of Greek thought:[12]

> The Christian-theological interpretation of the world that followed, as well as the later mathematical-technical thinking of the modern age, have removed science from its beginnings both in time and in its objects [*zeitlich und sachlich*]. For, assuming that the original Greek science is something great, then the *beginning* of this great thing remains its *greatest* moment. The essence of science could not even be emptied and used up—which it is today, all results and "international organizations" notwithstanding—if the greatness of the beginning did not *still* exist.[13]

To progress into the New German State, he, his faculty, and his students must go back to these beginnings. If one agrees with the claim that great things are greatest in the beginning, then to secure our future we must rediscover and reclaim the past. That philosophic task will help ground mere progress in knowledge upon the firmer foundations of valid philosophy: "Otherwise, science will remain something in which we become involved purely by chance or will remain a calm, pleasurable activity, an activity free of danger, which promotes the mere advancement of knowledge [*Kenntnisse*]."[14]

It is difficult to know why Heidegger denigrates scientific progress. Learned scholars have said and will say more about this aspect of his work. No doubt it is unfair to reduce Heidegger's mature thought to this one speech; nor is it fair to moralize about his actions from our time vastly removed from his situation in 1933. Yet, he reiterated the major claims of the rectorship address when he commented on the Greeks and Hegel in 1958:

> Within the title "Hegel and the Greeks" it is the whole of philosophy within its history that speaks, and that today in a times in which the collapse of philosophy becomes flagrant; because it has migrated into logistics, psychology, and sociology. These autonomous domains of research assure themselves of increasing importance and polymorphous influence as functional forms and performance instruments in the political-economic world, that is, in an essential sense, of the technical world.[15]

It is clear that Heidegger does not judge "mere advancement of knowledge" as the highest goal of the university. I find that a tragic error and I cannot make sense of the denigration of medicine, for example. Apart from the task of assessing why Heidegger wrote this way, I make the simple point that he used his immense talents to attack the very idea of progress and offered, instead, regression to idealization and spiritualism. To retrieve our spiritual essence, Heidegger sends us back to the Greeks. Obeying his injunction, we now go back to the Greeks.

PROGRESS IN GREEK PHILOSOPHY, LITERATURE, AND MATHEMATICS?

Our question is, again, why is it so difficult to denote progress in the humanities? There already exists a rich history of science in which the idea of progress is thoroughly investigated. Alongside T. S. Kuhn's celebrated monograph, *The Structure of Scientific Revolutions*, are numerous and subtle studies of progress in the sciences. If we read T. S. Kuhn and Karl Popper we learn that many of the problems that plague the humanities—the surplus of schools, the endless wrangling, the lack of definitions and measures, etc.—denote the state of a discipline just before it assumed the mantle of science.[16] According to these authors, those disciplines we call the humanities are merely protosciences; when they mature they will emulate the sciences. Thus, they will progress just as the sciences progress.

Looking back at numerous efforts to make the humanities scientific, we find smart people spending lots of energy with little effect. One way to review this struggle is to reflect upon the Greeks of the classical period, who also worried over this question. By 360 B.C.E. or so, Plato and other Greek thinkers looked *back* to those who had gone before them. Each tried to show how his new discipline, especially philosophy, was better than those of the earlier period. (In contrast to Heidegger, they felt confident that progress was possible.) We can appreciate these struggles by reviewing, briefly, classical Greek philosophy and comparing it to Greek art and literature and both to Greek mathematics.

— — —

Philosophic debates about progress in philosophy go back some twenty-five hundred years in the West, beginning at least with Plato. Plato, who was born around 428 B.C.E. and died in 348 B.C.E., explains how his teachings are advances over the poets and myth makers of the past. In *Phaedrus*, written around 360 B.C.E., he distinguishes poetry and myth (which are charming) from philosophy (true dialectic). The latter counts as progress. Socrates converses with Phaedrus, a young man who is enamored of the high-flown rhetoric of Sophists and others who use their verbal skills to persuade people to agree with whomever pays them the most. Phaedrus, who is a youth or *kourous*, has praised a speaker who argued that love is a kind of madness, catalogued according to its divine origins:

> The divine madness was subdivided into four kinds, prophetic, initiatory, poetic, erotic, having four gods presiding over them; the first was the inspiration of Apollo, the second that of Dionysus, the third that of the Muses, the fourth that of Aphrodite and Eros. In the description of the last kind of madness, which was also said to be the best, we spoke of the affection of love in a figure, into which we introduced a tolerably credible and possibly true though partly erring myth, which was also a hymn in honour of Love, who is your lord and also mine, Phaedrus, and the guardian of fair children, and to him we sung the hymn in measured and solemn strain. (265b)[17]

As charming as they are, Socrates rejects these myths of the origins of divine madness. The young man continues to praise the skills of rhetoricians, like Lysias and Thrasymachus, and their disquisitions upon the nature of love. Socrates refuses to accept this celebration of mere rhetoric. Like poetry, rhetoric can satisfy us emotionally, but it depends upon a slight of hand: it appeals to our feelings, it does not help us achieve a unified experience of soul. The only way to achieve that is through the hard work of dialectic.

If rhetoricians like Thrasymachus could, in fact, name the true parts of the psyche, Socrates would follow them. To think clearly we must be able to distinguish essential, that is, logical features, of concepts like *love*. However, entrepreneurial teachers of rhetoric, like Thrasymachus, fail to make logical distinctions:

> And those who have this art, I have hitherto been in the habit of calling dialecticians; but God knows whether the name is right or not. And I should like to know what name you would give to your or to Lysias' disciples, and whether this may not be that famous art of rhetoric which Thrasymachus and others teach and practise? Skilful speakers they are, and impart their skill to any who is willing to make kings of them and to bring gifts to them. (266b)

Yet, even great speakers like Pericles, taught by the esteemed thinker Anaxagoras, failed to ground their thinking in accord with the study of nature and the soul (271a).[18] In contrast to mere rhetoric, the *philosophic* task is twofold: to define the subject matter, in this case, love, and to distinguish types of love each based on some natural (or logical) form:

> The second principle is that of division into species according to the natural formation, where the joint is, not breaking any part as a bad carver might. (266a)

This lovely metaphor captures Plato's sense of nature; that it divides into natural parts, as the leg differs from the hip. The true philosopher should make that natural division the basis of his or her distinctions. Sophists, in contrast, compound false distinctions or even doublings to make their argument fit their needs; they will ascribe evil to one part, and beauty to another. By obeying their clients' wishes, Sophists play to their audiences' prejudices and feelings. They do not look to nature itself and thus they do not offer progress.

In the *Phaedo*, written around 360 B.C.E., Plato makes a similar distinction between *historia* and philosophy. Socrates describes his initial enthusiasm about doing *historia*, the investigation of the causes of generation and decay:

> When I was young . . . I was tremendously eager for the kind of wisdom which they call investigation of nature [*peri physeōs historian*]. I thought it was a glorious thing to know the causes of everything [*eidenai tas aitias hekastou*], why each thing comes into being and why it perishes and why it exists. (96a)[19]

Historia is derived from the noun *histōr*, a word found in several dialects of older Greek. Used both as a noun and as an adjective, it means "one who knows," "one who has seen," and "one who is acquainted with the facts." The *histōr* is not merely one who knows, but someone who puts this knowledge to effect. *Historeō*, the verb, means to know in the special sense of using knowledge effectively. Thus, *historia* came to mean knowledge gained through inves-tigation.[20] Practitioners of Ionian *historia* (Thales, Anaximander, and Hecataios of Miletus) reflect this general concern with knowing things about the world as a whole. A mark of this is the diverse topics they engaged, e.g., mathematics, meteorology, biology, history, mechanics, and cosmology.

Socrates finally rejected their approach because he found that even Anaxago-ras, who stated that mind (*nous*) was the cause of all things, provided no real explanation of the world order.[21] Anaxagoras "made no use of intelligence, and did not assign any real causes for the ordering of things, but mentioned as causes air and ether and water and many other absurdities" (*Phaedo* 98c). Plato is here criticizing the physicists' formal program, namely attempting to account for all phenomena in terms of (using W. K. C. Guthrie's phrase) "impersonal forces."[22]

Plato is also struggling to distinguish types of causes: the Greek term he uses in the sentence above is *aitias* from the noun *aitios*. From Liddell and Scott we learn that the first and most common meaning is "blameworthy" or "culpa-ble"; the secondary meaning is "cause of" in the sense of moral responsibility. In this sense, "real causes" must be more than accounts of physical actions; for real causes derive from an intelligence that drives those actions. It is those moral or intellectual causes that we should investigate: "If anyone were to say that I could not have done what I thought proper, if I had not bones and sinews and other things that I have, he would be right. But to say that those things are the cause of my doing what I do, and that I act with intelligence but not from the choice of what is best, would be an extremely careless way of talking" (98c).

In a moment of high drama, Socrates says that both the Athenians who judged him as impious and he who argued against them exemplify causal rea-soning. For both his accusers and he acted upon reasoning and the search for moral truth: "The real causes, which are, that the Athenians decided that it was best to condemn me, and therefore I have decided that it was best for me to sit here and that it is right for me to stay and undergo whatever penalty they order" (98e).

It is not just personal acts of intelligent beings that the physicists fail to address. They also fail to account for the orderly arrangement and workings of the cosmos: "They do not look for the power which causes things to be now placed as it is best for them to be placed, nor do they think it has any divine force, but they think they can find a new Atlas more powerful and more all-embracing than this" (99c). Socrates (Plato) objects most strenuously to the doctrines of Anaxagoras and others who conducted *historia* because they gave no real causes for the ordering of things.

Was Plato correct? Here we enter into a landscape well known to humanists: the realm of disputed interpretations. We find respected historians who disagree with Plato and who argue that Greek philosophy and Greek science (*historia*) are the same thing. For example, W. K. C. Guthrie says: "Philosophy is born when . . . for religious faith there is substituted the faith that was and remains the basis of scientific thought." One might argue against Guthrie and distinguish early Greek science from early Greek philosophy.[23] Early scientists like Thales, Anaximander, and Anaximenes used arguments to support their theories. And, beginning with Thales we find critical discussion. We need not deny that early Greek science was rational inquiry; we can deny that it was philosophic in the way that Plato defined philosophy. It was nonphilosophic because it was not conceptual inquiry. It was inquiry that sought to explain what the world was in terms of itself.[24]

If we accept Plato's distinction (that *historia* is distinct from philosophy in the way that legs differ from hips), Guthrie is wrong. For, again, those who practiced *historia* lacked what Plato called a concern for the causes of intelligent nature. According to this way of reading the history of Greek philosophy, true philosophy is concerned with the problem of understanding how and what we may know. Peter Winch makes a similar claim in *The Idea of a Social Science*, where he compares science and philosophy: "Whereas the scientist investigates the nature, causes and effects of particular real things and processes, the philosopher is concerned with the nature of reality as such and in general."[25] On the side of Plato and contra Guthrie is John Burnet. In *Greek Philosophy* Burnet notes that the question "What is real?" pertains to "the problem of man's relation to reality, which takes us beyond pure science it is not an empirical question at all, but a conceptual one."[26]

Plato makes this same point when Socrates says in the *Phaedo*:

> "After this, then," said he, "since I had given up investigating realities, I decided that I must be careful not to suffer the misfortune which happens to people who look at the sun and watch it during an eclipse. For some of them ruin their eyes unless they look at its image in water or something of the sort. I thought of that danger, and I was afraid my soul would be blinded if I looked at things with my eyes and tried to grasp them with my senses. So I thought I must have recourse to conceptions and examine in them the truth of realities [*skopein tōn ontōn tēn alētheian*]." (99e)

If we follow Plato, we may distinguish *historia* from philosophy on the basis of their respective concerns and methods of investigation.[27] A *historian*, like Anaximander for instance, attempts to explain the genesis and order of things in terms of wholly natural processes. He conceives of the world as a problem, a thing to be investigated. In contrast, a philosopher investigates the world by investigating the nature and logic of the conception. The first endeavor, that of the physicist, results in physical theory; the second, philosophical analysis, results in epistemology. Epistemological inquiry into the conditions of knowledge requires that one develop a metaphysics. Further, because metaphysics involves making claims about the nature of phenomena that are not merely natural, that is subject to *historia*, the metaphysician must employ logical argument.

Thus the first instance of doctrines that were contrary to common sense appeared in the thought of the first Greek metaphysician, Parmenides of Elea.[28] Some have argued that because he was the first metaphysician and because he provided a critique of commonsense conceptions, Parmenides provided Greek science with what it had lacked before, a paradigm of logical argument. Karl Popper makes a similar point when he traces the basis of modern scientific thought back to the geometric model of Plato's metaphysics: "What we find in Plato and his predecessors is the conscious construction and invention of a new approach toward the world and toward knowledge of the world. This approach transforms an originally theological idea, the idea of explaining the visible world by a postulated invisible world, into the fundamental instrument of theoretical science."[29]

Plato follows this speech with the famous metaphor of the cave: most humans live in a cave and do not know that what they think is real is but a series of shadows and images (*eidōlon*) cast on the cave walls. The goal of genuine philosophy is to see through the shadows and to ascend from the cave: "Then this is the progress which you call dialectic?" asks his interlocutor. Yes, says Socrates: "This power of elevating the highest principle in the soul to the contemplation of that which is best in existence, with which we may compare the raising of that faculty which is the very light of the body to the sight of that which is brightest in the material and visible world—this power is given, as I was saying, by all that study and pursuit of the arts which has been described."[30]

According to this reading of Plato, we can argue that his form of philosophy was better than both rhetoric and *historia*; the first did not attend to moral reasoning and the role of mind, the second did not attend to the role of logic and attention to conceptual principles. While I happen to be persuaded, we note that W. K. C. Guthrie, the Laurence Professor of Ancient Philosophy at Cambridge University and author of six massive volumes on the history of Greek philosophy, is not. A betting person should wager on the Laurence Professor of Ancient Philosophy and not on me, except that as I read Plato he's on my side. Of course, Professor Guthrie has written deeply and at length on Plato, and we could, if we wished, find in those volumes an eloquent defense of his reading of Plato.

This brings us back to B. F. Skinner's lament: that some read the Greeks as if they told us something valuable about human experience. Skinner's lament should start with a more elementary fact: we do not begin with the same reading of Plato. For it is not as if we agreed upon what Plato meant or what Plato said. On the contrary, we often begin with contested interpretations of Plato just as Plato struggled against the pre-Socratics and as Aristotle struggled against Plato. (A similar set of contested interpretations dominates discussions of David Hume, the English skeptic, Immanuel Kant, Friedrich Nietzsche, Martin Heidegger, Karl Barth, and any other thinker who merits the title "great philosopher" or "great theologian.")

DEVELOPMENT AND PROGRESS IN GREEK SCULPTURE?

Art historians and archaeologists can trace a developmental line that stretches from Egyptian sculpture through the archaic age of Greek arts (ca. 600 B.C.E.) through the classical period (the height of Athens's power around 420 B.C.E.) to the Hellenistic period (a hundred years later). It is tempting to call this progress.[31]

Art historians can show that these *kouroi*, images of young men like Phaedrus for example, developed over a five-hundred-year period. However, proving that later pieces are *better* than earlier pieces is not easy. It seems true that the archaic pieces are rigid and less lifelike than the classical pieces; does that make the archaic lesser? If we agree upon objective criteria, then we can agree that the later pieces count as progress over the earlier pieces. We need, though, to first agree upon what counts as objective criteria.

F. B. Tarbell, a classics professor at the University of Chicago around the beginning of the twentieth century, offered one such list. In his *History of Greek Art* (1896), Tarbell provides a standard account of the development of *kouroi*.[32] Thus we read in chapter 6, "The Archaic Period of Greek Sculpture, Second Half 550–480 B.C.," that the sculpture of this period was better than that which preceded it but not as good as that which emerged in the classical period about a hundred years later. To justify these judgments of progress, Tarbell produces a lengthy account of how the later pieces improved upon the earlier ones:

> True, many traces still remain of the sculptor's imperfect mastery. He cannot pose his figures in perfectly easy attitudes, not even in reliefs, where the problem is easier than

in sculpture in the round. His knowledge of human anatomy . . . is still defective, and his means of expression are still imperfect. For example, in the nude male figure the hips continue to be too narrow for the shoulders, and the abdomen too flat. The facial peculiarities . . . in the preceding chapter—prominent eyeballs, cheeks, and chin, and smiling mouth—are only very gradually modified. As from the first, the upper eyelid does not overlap the lower eyelid at the outer corner, as truth, or rather appearance, requires, and in relief sculpture the eye of a face in profile is rendered as in front view.[33]

Like other commentators upon the "Greek miracle," Tarbell focuses upon the speed and direction of change in Greek sculpture that we can see occur within a few hundred years. As Tarbell and others have noted, they differ dramatically from one another along the dimensions that Tarbell lays out. We can compare "Kouros from Cape Sounion" around 600 B.C.E., which is more formal and less rounded than a second "Kroisos" made some sixty years later. Later statues of Zeus and Herakles are, as Tarbell says, anatomically richer, even if more idealized than those that preceded them. Finally, later pieces, such as "Sleeping Satyr," appear one hundred years later and in their plasticity and brilliance continue the same virtues noted in the classical period.

Are they therefore better? Here, again, we enter the jostle of disputes over the meaning of "best" according to the dictates of one's dominant conception of what counts as a fully mature human being. If one believes that the refined surface of the Sleeping Satyr, its staging, its exacting anatomical detail, and its posture represent something new and true, especially about male eroticism, then the latest in this group of five is also the greatest. But not everyone believes this. Like Heidegger, a large number of European and American thinkers saw in Hellenism a decline from the heights of the classical (and preclassical) period. While they acknowledged the technical mastery of the Hellenistic works—and their outstanding emotionalism and power—they saw these works as declinations away from the greatness of the earlier period.

Nigel Spivey notes that earlier German commentators upon Greek statues, like J. J. Winckelmann (1717–1768), claimed that the greatness of the classical period's sculptures lay in their origins in Greek democracy of the same period. According to this political reading of art, when Athenian democracy faded, so too its art faded. Wicklemann's desire to correlate great art with the great civic value of liberty floundered when he admired pieces done long past the fifth century B.C.E. when Athenians ruled themselves. Spivey summarizes the mental gymnastics that Winckelmann had to perform when he praised a portrait of the boyfriend of Emperor Hadrian (117–138 C.E.). To justify his praise of the portrait and preserve his linkage of great art with liberty, Winckelmann claimed that when this portrait appeared, "Hadrian was 'planning to restore to the Greeks their original freedom, and had begun by declaring Greece to be free.'"[34]

As an art historian, Spivey correctly notes that this wish is countered everywhere by the facts: great art emerged under tyrants, under benign princes, and sometimes under the rule of the *dēmos*. Winckelmann's wish to show that great art must emerge from superior political structures is not nonsensical. It takes us back to Heidegger's address; both men yearned to connect what is highest in one sphere of human endeavor with what is highest in another. Spivey cites W. B. Yeats: "Europe was not born when Greek galleys defeated the Persian hordes at Salamis, but when the Doric studios sent out those broad backed marble statues against the multiform, vague, expressive Asiatic sea."[35] When Yeats, a great Irish poet and nationalist, joined a committee to oversee the design of Ireland's new coins in 1928, he encouraged them to follow the example of ancient Greek coins because "Ancient Greek coins . . . were statements of their independent status and an enduring expression of identity."[36]

The urge to distinguish classical Greece and its democratic values from the "Asian hordes" appears throughout English commentaries on Greek statues. Thus, in his famous *Greek Studies*, Walter Pater contrasts the Greek search for truth against the Egyptian love of rigidity and predictable form:

> In representing the human figure, Egyptian art had held by mathematical or mechanical proportions exclusively. The Greek apprehends of it, as the main truth, that it is a living organism, with freedom of movement, and hence the infinite possibilities of motion, and of expression by motion, with which the imagination credits the higher sort of Greek sculpture; while the figures of Egyptian art, graceful as they often are, seem absolutely incapable of any motion or gesture, other than the one actually designed. The work of the Greek sculptor, together with its more real anatomy, becomes full also of human soul.[37]

Given a value system not wedded to political ideology, one might argue that even pieces from the fifth century illustrate a decline from the spirit of the earlier, perhaps naive, but fresher period. Thus, in a recent book, Mary Stieber says that classical art can easily fall into mere academicism, into schools and mere technique. Archaic art, on the other hand, "at its highest levels of execution, is capable, in its way, like the best Analytical Cubism of Picasso and Braque, of attaining realism in spite of its less than perfect naturalism."[38] Stieber's thoughtful essay dissolves the wish to correlate "later art" with "better art" and realism with naturalism. Each of these concepts reverberates in the history of art and art criticism; none provides a conclusive way to establish progress. We take up this issue again and look at another aspect of Greek culture, its literature.

GREEK LITERATURE,
"MORE SERIOUS THAN HISTORY"

The difference between a historian and a poet is not that the one writes
in prose and the other in verse. . . . The real difference is this, that one
tells what happened and the other what might happen. For this reason
poetry is something more scientific and serious than history, because poetry
tends to give general truths while history gives particular facts. (Aristotle,
Poetics 1451b)[39]

In our time, few would assert that poets are more scientific than historians. Giv-
en the gulf that separates the sciences from the arts and the humanities, a predictable bifurcation ensues. History and its sibling disciplines in the social sciences attempt to become scientific by following (or mimicking) the trappings
and methods of the natural sciences. The pursuit of this title is understandable.
The term *science* has acquired a prestige analogous to that of faith in the Christian middle ages. Those who support, affirm, and advance science are on the
side of progress; those who do not, flirt with irrelevance, if not heresy. But Aristotle says that poets are more scientific than historians and, by extension, all
those who merely assemble facts.

It seems that the translator has slipped one over on us. For the Greek text
does not talk about science as *epistēmē* but of *philosophia*, which looks a lot like
philosophy. Examining Aristotle's use of the term we learn that *philosophia*
means the methodological examination of the world. It denotes the knowledge
of the perceptible reality of the kosmos.[40] In *Poetics* as in *Metaphysics* and other
treatises on first principles, Aristotle uses *philosophia* interchangeably with
epistēmē; the latter denotes true knowledge and understanding. Aristotle says
that poets grasp something true about the patterns that make up human lives,
even if their constructs, their narrations, are not historical recounting. In fact,
he says that the mere accretion of facts does not constitute poetry and cannot
constitute science.

The poet's insights into what might happen and must inevitably happen refer
to consequences of human choices and the effect of character. To cite another
famous Greek saying, "Character is fate." Poets and other artists who comprehend patterns of human motivation grasp something true about future behavior
and therefore future events in a life story. Given the Greek philosophic sense of
Justice (*Dikē*), Necessity (*Anankē*), and Natural Law (*Themis*), to grasp the pattern of human character, especially in its hidden aspects, is to grasp something
true about the cosmos itself. For Justice, Necessity, and Natural Law all pertain

to the overarching rules that govern the gods as well as humans, divine actions as well as mundane actions. This deeply shared affinity between humans and the gods demarcates human being:

> Zeus himself ordained law for mankind. As for fishes and beasts and winged fowls, they may feed on one another without sin, for justice is unknown to them. But to man he gave the law of justice. (Hesiod, *Works and Days* 276–79)[41]

This passage is more religious than Aristotle's reflections, since it emphasizes the creaturehood of human beings. Yet, it asserts a customarily Greek value: that the human mind is itself a cosmos, a unified entity that reflects the unity and coherence of the universe. In hundreds of distinct passages, the pre-Socratics, Plato, and Aristotle wrangle over the implications of this theorem of unity. For the very idea of Metaphysics, a First Philosophy that addresses the nature of Being as such, presupposes that there is a unity to being and that the human mind can grasp this unity.

This claim may appear obvious to us. It is obvious only because it underlies the assumptions of science as we know it in the West. That science in all its forms is at all possible fails to stun us only because we automatically share Aristotle's project. We share the Greek understanding of human fate and the possibility of comprehending its patterns: "Divine activity neither controls human activity and suffering nor renders them merely pathetic, but is rather a generalized statement about them. The divine background holds up to us, so to speak, the system of co-ordinates against which we are to read the significance of what the human actors do and suffer."[42]

The claim of the unity of being itself did not echo in popular Greek psychology. For within the Greek world and perhaps within all folk cultures, the philosophic claims about the nature of being, especially its unity and eventual discoverability, did not expunge affirmations about the supernatural. While Greek metaphysicians asserted a fundamental unity between human character, the psychic cosmos, and the external cosmos, Greek poets affirmed doctrines that are inherently dualistic:

> Although the body of every mortal must give way to overmastering death, yet an image of life lives on, for it alone comes from the gods. It sleeps while the limbs are alive, yet even during sleep it gives to men in many a dream premonitions of happy and disastrous turning-points. (Pindar, *Threnodies* 131)[43]

This fragment from Pindar's dirge suggests that a human being is of two natures: the mortal stuff that fades away and a something else, an image (*eidōlon*) of life that derives from the immortal gods. For this reason, Christian theologians could cite these lines in their essays on the divine origins of the soul.[44] This second substance or entity grants to humans occult powers made evident in extrasensory experiences of dreaming or during other altered states of consciousness. For in those states we cannot distinguish real and imaginary since our sense of reality during that state of consciousness proclaims that this also is real.

This conceptual tension between a philosophic monism and popular dualisms is understandable. Regarding conceptual problems that do not yield a single answer, human beings typically adopt a dualistic response. Are the body and mind of one or two substances? Is Human Being identical to God or separate? Answering either yes or no to such questions evokes a persuasive rebuttal from the other side. Hence, wise persons, like Aristotle, typically affirm both answers. This makes them vulnerable to critics. For using Aristotle's own logic, to affirm both yes and no to such propositions violates the law of the excluded middle: therefore his propositions are false. Tragic poets are willing to tolerate these conflicts. They bring contradictions within a culture's self-understandings into the light. Aristotle celebrated Greek drama because it exemplified that form of courage. In this sense, we can trace a progressive line that stretches from Greek myth to Greek theater but we cannot call it scientific.

ANTIGONE, OEDIPUS THE KING, OEDIPUS AT COLONUS

In *Antigone*, the third part of Sophocles' trilogy, we tumble headlong into an irresolvable conflict. Obeying the dictates of her religion, Antigone seeks to bury her dead brother, Polyneices, who has led a rebellion against Creon, the new king. Creon has issued an edict forbidding any honor to a rebel. Creon speaks of duty, historical precedent, good government, and revenge upon traitors; Antigone speaks of filial love, family duty, and the gods' unwritten and unfailing laws (499–524).[45] These divine laws override those of mere mortals, including kings. While Creon has declared death to anyone who disobeys, Antigone disdains that threat since she, like all sensible persons, knows that death will come to her someday: "And if I am to die before my time I consider that a gain" (515–16). Far worse would it have been to let her mother's son lay unburied, rotting on the ground.

These truths, Antigone says, are antithetical to each other. She cannot imagine a synthesis or an accommodation with Creon: "Your moralizing repels me, every word you say—pray god it always will. So naturally all I say repels you too" (557–59). This latter sentiment solidifies Antigone's character and reveals, at the same moment, her inherent limitations. For from her single-mindedness, her intense devotion to the duty toward blood, can come no harmonious resolution of the culture's dilemma—how to adjudicate between two competing ethical and religious demands.

Compared to the more rounded figures of later European theater, especially to Shakespeare's characters, Antigone and Creon, and even Oedipus, are stark. David Grene noted that within the trilogy, which is founded upon a myth, one can see that Creon, whom moderns might see as a Nazi despot, actually occupies the same thematic position as does Oedipus. In the central Greek struggle over the nature of the state and leadership, both kings represent an aspect of kingship: "The tyrant who with true and good intentions orders what is wrong, morally and religiously, is crudely represented in Creon; he is much more subtly represented in Oedipus himself."[46] This generic side of tragedy, as Grene calls it, parallels what Aristotle termed its scientific nature; it represents the consequences of violating inexorable laws.

Moving and grand is Antigone's focus on duty. We would not wish her to follow her filial duty blindly without self-awareness. The nobility of her sacrifice is measured only by the depth of her self-understanding. Her sacrifice cannot be heroic if it is automatic. To clarify and reaffirm the transcendent value of filial piety and religious duty, Antigone must die. Once she has grasped this duty her character and its nobility is revealed and her fate is sealed; Creon will destroy her. But, of course, once Creon carries out that action his character is revealed, his fate and the fate of his kingdom is sealed.

In Aristotle's favorite play, *Oedipus the King*, the first part of the trilogy but the last written, it is Oedipus's overweening pride (his narcissism we would say) in his great intellect that requires him to fall victim to the oedipal curse. True, the folktale sets a context of fate in which Oedipus and others are victims of an ancient prediction and ancient bloodguilt. But neither Sophocles nor Aristotle (nor Freud) relieves Oedipus of complete responsibility for his actions; that would lessen his stature in our eyes and diminish the clarity of the play's insights. The inevitability of character makes possible poetic insight, and this justifies our calling poetry, especially when at the level of tragic theater, scientific.

This sense of inevitability also underlies our ability to be moved by the play and to imagine ourselves caught up in the hero's conflicts. For like him we understand dimly that our character also determines our fate, even if we vehemently blame our parents, siblings, social-cultural mores, and other external forces. Just as Aristotle saw himself as discovering general patterns within nature, so too the artist, like a great playwright, discovers patterns within human lives and human motivations. Thus while we may wish to be surprised by the turn of events within a story, we also wish to feel that the story as a whole is an entity complete in itself. In Aristotle's famous phrase, we wish to comprehend an internal necessity that links the story's beginning, middle, and end. This internal coherence of the story is a mark of its excellence and why Aristotle assigns superiority to tragic drama.

Given this evaluation the poet has an interesting task. She must entrance us with something new and unexpected but at the end reveal a hidden structure that once seen will yield to us a full measure of aesthetic understanding. Or, in the language of *Poetics*, good poetry evokes in us amazement, fear, and pity. For some contemporary critics this criterion is too close to melodrama, show tunes, and popular fiction. Aristotle wants art to entrance him, to remove him from his ordinary consciousness and scientific reasoning, capacities he seems to have had in abundance. We can follow Aristotle to the theater, but we cannot agree that poetry advances in the way we see typical of the sciences. The sciences resolve conceptual tensions; the arts amplify them. Science reduces mysteries to puzzles and then solves the puzzle; art raises the mundane to the mysterious and draws us in. To get lost in a play, or a poem, or a dance, or a piece of music is to be changed, somehow, and we acknowledge that change later through paradoxical language. Lacking that, we cite another instance of art, such as Wallace Stevens's celebrated poem "Anecdote of the Jar," written in 1919:

> I placed a jar in Tennessee,
> And round it was, upon a hill.
> It made the slovenly wilderness
> Surround that hill.[47]

We need both scientific and artistic forms of thought, as Aristotle said, but the histories of the two are distinct, as we see when we consider Greek mathematics.

PROGRESS IN GREEK MATHEMATICS: INCOMMENSURABILITY

By the fifth century B.C.E.—the so-called classical period—Greek mathematicians had discovered that the magnitude of certain line segments, like the edge and diagonal of a square, are not commensurable with one another. This means that there is no rational number, a number that can be expressed as the ratio of two integers, which can designate this magnitude. How they discovered the existence of these nonrational numbers is a fascinating story and, like many issues in the history of Greek thought, a disputed one. Two magnitudes are considered "commensurable" (*symmetra*) with one another if there is a measure that divides each magnitude an exact number of times, designated by an integer (*arithmos*). If there is no such integer then the two magnitudes are considered incommensurable (*asymmetra*). For example, in a square with one-unit sides, the diagonal's length is the square root of the sum of the two sides: "Today we simply say that the diagonal's length is $\sqrt{2}$ (an irrational number). However, for Greek mathematicians the discovery of incommensurability meant that there existed geometric magnitudes that could not be measured by numbers. For they viewed numbers as integers alone, whereas the phenomenon of incommensurable lengths implied that geometric magnitudes have some sort of inherently (and unavoidable) continuous character."[48]

Greek mathematicians solved their dilemma by refusing to term the length of the diagonal a number: "Whereas such a simple equation as "x squared = 2" has no solution in the domain of (Greek, rational) numbers, the equation "X squared = ab where a and b are given lengths can be solved geometrically by constructing a square with edge x whose area is equal to that of the rectangle with sides a and b." In other words, while they could not designate an irrational number, Greek mathematicians could designate a quantity, namely an area, which correlated with that number. As taught in schoolrooms around the world, this is the Pythagorean Theorem. Area C^2 is equal to the sum of area A^2 and area B^2. When A and B are of equal lengths, Greek mathematics could not name the length of C using a rational number.

This geometric solution to the problem of irrational numbers meant that the Greeks had to tolerate a conceptual gulf that separated arithmetic from geometry. This gulf was not bridged until the modern period, when Leibniz and Newton separately developed methods and notations that made possible the modern calculus. (The latter is an instance of progress in a discipline that is neither a natural science nor a humanities.)

Historians of Greek science and mathematics note that in their struggles with the problem of incommensurables, Greek geometers proposed diverse responses; some were pragmatic, others were conceptual—none of them counted as full-fledged solutions. Pragmatically, some Greek mathematicians refined the Egyptian and Babylonian practices of approximating quantities that they could not calculate directly. A common example is "squaring the circle," that is, approximating the area of a circle by imagining an ever-larger number of polygons inscribed within it. By finding the area of this set of polygons, using basic geometry, we can approximate the area of the circle that they "fill up."

Conceptually Greek thinkers attempted to unite mathematics with geometry by way of the notion of the infinitesimal. Democritus suggested the infinitesimal was a monad, or "unit of such a nature that an indefinite number of them will be required for the diagonal and for the side of the square."[49] The notion of an infinitely small unit saves the concept of *ratios*, for it permits us to say that the ratio of the diagonal to the side of a square (which we now designate $\sqrt{2} : 1$) is a ratio of one set of infinitesimal units to another set of such units.

Zeno, a student of Parmenides, countered this solution. Aristotle attributes to Zeno a brilliant polemic against Democritus's concept of the *infinitesimal*: "That which, being added to another does not make it greater, and being taken away from another does not make it less, is nothing."[50] In addition, Zeno elaborated his famous paradoxes. As a set, they demonstrate the implausibility that there could be a class of infinite temporal or spatial units. For example, in his Achilles paradox Zeno showed that if a tortoise had a head start in a race with a superior runner, like the great athlete Achilles, then Achilles would have to cover that distance before he overtook the tortoise. If between the tortoise and Achilles are infinitely many spatial units, each of which he must traverse, then Achilles would require an infinite length of time to catch up. This he could never do. Refinements of this paradox show that the tortoise could never traverse the infinitely divisible space stretching out in front of him. Indeed, if space and time are infinitely divisible, then all motion is impossible. Since this is an intolerable conclusion, Zeno rejects the concepts of infinitesimal units of space and time.

A satisfactory answer to Zeno emerged with calculus, the quantitative language of Western science. First, by rejecting the notion of infinitely divisible "real space" and "real time," we reject Democritus's amalgamation of mathematical entities with actual entities. Second, the concept of an infinite set of rational numbers lying, for example, between integer 1 and integer 2, is easily demonstrated to be a mathematic truth. It has no immediate claim upon natural space and time that we arbitrarily measure with fixed units of distance, like the marks on a piece of wood called a ruler.

What was difficult for minds as refined as those of Greek geometers is simple for us, who come after the eighteenth century. Either we are smarter than Democritus and Aristotle, for example, or we rely upon concepts not available to them. The evidence suggests that the second is more likely. One can read Aristotle's struggle with Zeno's paradoxes as the result of Aristotle's commonsense metaphysics. This rests upon "the cardinal weakness of Greek logic and geometry: naive realism which regarded thought as a true copy of the external world."[51]

Naive realism seems eminently plausible since so much of Euclid's *Elements*, for example, seem not just logically consistent, but accurate descriptions of the world we experience in our daily life. One can sympathize with Aristotle's wish for a unified metaphysics in which all the sciences, including geometry, had a fixed place within a hierarchy of thought. That this hierarchy matched perfectly the naive Greek sense of the world only added luster. Further, if with dialectic, one could show the validity of syllogistic reasoning, and if one began with valid assumptions, then we could deduce new truths about the world through reasoning alone.

In contrast to the wish for a unified metaphysics, Archimedes, usually denoted as the greatest of all Greek mathematicians, perceived that fundamental gaps persisted in geometry. These gaps became evident to him when he confronted central tasks, like "squaring of the circle," and other problems that require the concept of real numbers, like $\sqrt{2}$. Acknowledging these gaps, Archimedes developed heuristic methods to support, not prove, his conclusions. One method we have seen: the method of "exhaustion." Another method was the use of reductio ad absurdum arguments. For example, by showing that the sum of approximations could not be more than or less than the figure arrived at, Archimedes supported the validity of the figure itself. These methods, while brilliant, afforded only indirect proofs. To methods derived from previous geometers, Archimedes added his insight into mechanics, the law of the lever.

By combining these pragmatic methods, Archimedes discovered techniques adequate to find the center of gravity of semicircle parabolas and other geometric figures. Historians of Greek mathematics conclude that even these solutions were inadequate because they resulted from the ban on discourse about the infinite: "The reasons for this ban are obvious: intuition could at that time afford no clear picture of it, and it had as yet no logical basis."[52] On emotional grounds Aristotle's solution to the problems posed by Zeno seemed satisfactory.

Yet, on logical grounds, it was not. Greek mathematics progressed in numerous areas, but not in all. This acknowledgement, that gaps persisted, is a measure of its scientific status. In contrast to Heidegger's wish to return to Greek metaphysics, seventeenth-century mathematicians sought to improve upon Greek mathematics. They did so by creating calculus, which is superior to those methods that preceded it. This brings us back to the question of progress in the humanities. For it is not just Freudians and Heideggerians who yearn to go back to the Greeks or some other idealized people. We share their wish to believe that the mind and the world are eventually commensurate with one another, that we can understand the problem of Being, as Heidegger put it, through thought alone.

It also represents a wish to unite ourselves with these esteemed ancestors whom we place at the beginning, in the Golden Age, when philosophy was first born—forgetting that it's impossible to clarify just what that sentence means. This wish also animates thinkers like Freud, who counted his discipline among the sciences, yet who felt called upon to challenge the reign of religion. We see this wish blossom in Freud's 1920 essay *Beyond the Pleasure Principle*: "Our speculations have suggested that Eros operates from the beginning of life and appears as a 'life instinct' in opposition to the 'death instinct' which was brought into being by the coming to life of inorganic substance. These speculations seek to solve the riddle of life by supposing that these two instincts were struggling with each other from the very first" (Standard Edition 18.61).

To solve the riddle of life, Freud echoes speculations announced by another ancient Greek, Empedocles. Here is a typical verse:

> By earth (he says) we earth perceive,
> by water water,
> By air bright air, by fire consuming fire,
> Love too by love, and strife by grievous strife.[53]

These two authors are separated by twenty-four hundred years. Of Empedocles we know that he came from a wealthy family, practiced medicine, and wrote at least two books, one titled *On Nature*, a book of science, the other called *Purifications*, which reflected in a "more rhapsodical and religious character" upon human nature.[54] It is surprising to see Freud, an heir to the sciences, revert to a style of thought and language that so closely parallels that of Empedocles. Why would Freud do this? Seventeen years after he wrote these lines in *Beyond the Pleasure Principle*, Freud cited Empedocles directly in "Analysis Terminable and Interminable" (1937). There he correctly notes how similar are his notions of two types of instinct to Empedocles' concepts of the powers of love (*philia*) and strife (*neikos*). They are, Freud says, "both in name and function, the same as our two primal instincts, Eros and destructiveness, the first of which endeavours to combine what exists into ever greater unities, while the second endeavours to dissolve those combinations and to destroy the structures to which they have given rise" (Standard Edition 23.246).

Concerned with the originality of his thought, Freud hastens to add that he had not known of Empedocles when he wrote *Beyond the Pleasure Principle*. Or, if he had known, he had forgotten. In either case, "I am ready to give up the prestige of originality for the sake of such a confirmation" (Standard Edition 23.245). It is striking that Freud would consider a poem by an ancient Greek writer a confirmation of his twentieth-century scientific proposition. There are many poems by many ancient Greeks whose opinions are at odds with one another. They cannot all, therefore, be true. To cite these verses as confirmation of his theory of the duality of human nature is to misuse the concept of *confirmation*. Unless, that is, one believes that progress in understanding human nature must take us back to our past, to our ancestors. Back to the Greeks! as Heidegger thundered. I have tried to show that this is an error based on our illusory wishes to have discovered the foundations of the "West," that is, of ourselves.

In their philosophical works, Greek masters *seem* to have laid out the principal problems for Western thought. In their arts, especially tragedy, they *seem* to have surveyed our basest and highest needs. Their self-portraits, their self-representations *seem* complete. After all, they are products of the *classical period*, a label that declares their superiority, their closeness, and, paradoxically, their distance from us who cannot join them. But this is an illusion. By idealizing the Greeks, first the Romans, then Renaissance potentates, then the English and other secular powers claimed a spiritual continuity with a Golden Age, with the beginnings. As I noted earlier, Heidegger claimed that "assuming that the original Greek science is something great, then the *beginning* of this great thing remains its *greatest* moment." This is false in at least two ways: first, many great things, like the *Sequoia sempervirens* (the California Redwood), cities, and universities, become greater over time. Second, the origins of a thing are not equivalent to current causes. Freud's discoveries about childhood sexuality (and the ubiquity of sexual contact in childhood) led him to say that adult neuroses were caused by traumata in early years. But this is not a valid conclusion: many children with intense sexual experience do not become neurotic; and many neurotics, Freud discovered, fabricate stories of early sexual experience. Attempting to remedy his earlier simplifications about how personal histories "caused" adult disorders, Freud created the concept of *transference*.[55] Transference names the fact that we bring to current relationships feelings and expectations shaped by how we currently view past relationships. Each of these factors is a current cause of my current behavior; none is a mere memory of events that happened long ago.

In contrast to the rounded, complete portraits of Greek art, Greek mathematics found no resolution to its central dilemmas until the seventeenth century. While mathematicians tolerate the intellectual gaps and incommensurability of reasoning, most of us cannot tolerate intellectual gaps and incommensurability about "human being." We demand more immediate answers, if not from the faltering sciences, then from science fiction and other arts. As we will see in the next chapter, in a curious way, by pushing themselves to formulate scientific diagnoses, American psychiatrists have, sometimes, told stories that look a lot like those found in *Star Trek*.

4. SEVEN OF NINE AND FIVE OF NINE

SCIENCE FICTION AND PSYCHIATRY

The good clinician, like the good pastor or rabbi of religious traditions, learns quickly that to help people get better we must link them to their and our myths, that is, to *our* narratives. For American psychiatrists and many psychotherapists, this means we should use the "five-of-nine" rule and similar rules announced in the *DSM-IV*, the bible of psychiatry. However, by doing so we find ourselves uncannily associated with science fiction and religion and their vast efforts to solve the problem of human being. Neither science fiction authors nor we can wait for rigorous science to solve questions that arise now. We cannot ask children (or adults) to wait a few hundred years until neurobiologists give us definitive answers about the validity of recovered memories or the plausibility of belief in free will. Suffering occurs in real time; if we cannot imagine it ending, the only way left is depression. When depression worsens, death looms as a welcome relief. Recognizing this sequence, religious geniuses tell stories that resolve ambiguities; what was unclear is made clear, what was hidden is revealed, if not now, then at the end of time. Known since the beginning of time, we learn of our place in life, our union with the foundation of being, with the Being behind being, as Paul Tillich put it.

The problem of defining personhood pervades religious and ethical reflection. The struggle to defeat slavery, for example, required ferocious attack upon claims of African inferiority and a counter-story about the transcendent unity of all human beings. In the brilliant motto of the Quaker antislavery campaign, represented everywhere in its medallions struck by Josiah Wedgewood, was an African in chains who asks, "Am I not a man and a brother?" This question puts the metaphysical burden upon the defenders of slavery. Unless they can show— scientifically—that Africans are *essentially* distinct from Caucasians, slavery cannot be justified upon racial grounds.

While the Quakers could not prove the contrary, that Africans were identical to proper Englishmen, they could rely upon the totalizing narrative of Jewish and Christian scriptures. In its austere stories, Judaism and Christianity addressed the question of human being: Who are we? And they answered: Children of God. Thus it follows, as the Quakers said, that because we share the same Father, Africans must be our brothers and sisters.

Where religious discourse stopped, Freud and twenty-first-century psychiatry and science fiction authors began. Freud and his uncanny others—novelists, playwrights, and science fiction authors—begin with life at the surface and plunge into the depths, seeking to find as-yet-unseen truths. They cannot begin with the directorial mastery of Genesis or Romans. This leaves them and us unsure about life at the boundary.

MAPPING THE BOUNDARIES OF HUMAN BEING

It may seem unfair to psychiatry to associate it with science fiction. Yet, the clinical discipline of psychiatry and the genre of science fiction (and its cousins such as fantasy, myth, and legend) share a common quest: to map the limits to human being and human consciousness. What is unclear and invisible in the world of psychiatry—Is the patient's anger justified or not justified? Did the therapist's empathy cause the patient's illness to disappear, or did the symptoms merely go into remission?—becomes clear and visible in the worlds of science fiction. For in those worlds, in *Star Trek* (and *Star Wars* and *The Lord of the Rings*), the invisible parts of the psyche, the parts at whose workings scientific psychiatry must guess, emerge as distinct characters. Psychiatrists use handbooks of symptoms, including checklists that require, for example, that a patient manifest five of nine behaviors to be labeled "borderline." Science fiction authors create characters that explore variations on the topic "human being." By concocting new worlds, with new challenges, the unthought, unknown parts of

Figure 4.1. Young woman.

ourselves emerge like a Polaroid picture emerges from the seemingly empty film torn out of the camera.

For example, Seven of Nine (who looks like the woman in fig. 4.1) is part human and part Borg (a machinelike superrace) and so must learn consciously what we learn unconsciously. Seven, as she is known to her friends, appeared in the American television show *Star Trek: Voyager*, broadcast on United Paramount Network (UPN) from 1995 through 2001. *Voyager*, of course, is one of six *Star Trek* television series, and they are but a fraction of the books, magazines, videos, DVDs, full-length films, animation series, conventions, and vast library of fan-fiction devoted to the fantasy world created by Gene Roddenberry forty years ago.

Figure 4.2. Cover of *Diagnostic and Statistical Manual of Mental Disorders*,
text revision, 4th edition (American Psychiatric Association, 2000).

The five-of-nine phenomenon is less exciting; it refers to the unhappy fact
that one must make psychiatric diagnoses using a list of symptoms in the fourth
edition of *Diagnostic and Statistical Manual* (*DSM-IV-TR*):

If a patient manifests five of nine behaviors (symptoms) denoted in the *DSM-
IV* associated with the term *borderline personality*, then that person merits the
label *borderline*.[1] This five-of-nine rule demarcates the compromises forced
upon clinical researchers and psychiatrists who strive to make rigorous, scientifically valid diagnoses by categorizing behaviors. The logic of this rule, that we
can choose any five of the nine symptoms listed, tells us that the entity in question is not like other diseases, including psychiatric diseases associated with discernable causes.

While this may seem to vitiate the *DSM-IV* enterprise, this judgment is unwarranted. Just because it's difficult to make diagnoses rigorous does not mean that no such things as personality disorders exist. Whole platoons of psychiatrists and psychologists have refined these categories to levels where, with careful training, one can make judgments that are reliable (that is, we make the same designation consistently) and valid (that one can designate this syndrome as distinct from other syndromes).

For many good reasons clinicians need to know that when researchers in Nashville investigate "borderline personality" they are referring to the same thing as researchers in Seattle or in Tokyo. I cite the problem of borderline diagnoses because it is among the most disputed; like other forms of personality disorder, it appears to be the most culturally dependent of the psychiatric categories. As diagnoses move away from categorizing behaviors and toward physical findings, such as "Mental Disorders Due to a General Medical Condition," their rigor increases and their contentiousness decreases.

Criteria for Borderline Personality Disorder (DSM-IV 654)

A pervasive pattern of instability of interpersonal relationships, self-image, and affects, and marked impulsivity beginning by early adulthood and present in a variety of contexts, as indicated by five (or more) of the following:

1. frantic efforts to avoid real or imagined abandonment. Note: Do not include suicidal or self-mutilating behavior covered in Criterion 5.
2. a pattern of unstable and intense interpersonal relationships characterized by alternating between extremes of idealization and devaluation
3. identity disturbance: markedly and persistently unstable self-image or sense of self
4. impulsivity in at least two areas that are potentially self-damaging (e.g., spending, sex, substance abuse, reckless driving, binge eating). Note: Do not include suicidal or self-mutilating behavior covered in Criterion 5.
5. recurrent suicidal behavior, gestures, or threats, or self-mutilating behavior
6. affective instability due to a marked reactivity of mood (e.g., intense episodic dysphoria, irritability, or anxiety usually lasting a few hours and only rarely more than a few days)
7. chronic feelings of emptiness
8. inappropriate, intense anger or difficulty controlling anger (e.g., frequent displays of temper, constant anger, recurrent physical fights)
9. transient, stress-related paranoid ideation or severe dissociative symptoms.

As critics have noted, these criteria are a mixed lot, both logically and ontologically. The first criterion alludes to interior *reasons* for specific actions ("frantic efforts") and categorizes these reasons ("to avoid real or imagined abandonment") as *causes* of those actions. The second invokes the concept of *pattern* of relationships and the equally complex concepts of *idealization* and *devaluation*. The third criterion depends upon the psychodynamic *concepts* of identity and identity disturbance and, again, invokes a criterion of stability and pattern: the swing in self-understanding must be *persistent*.

The fourth takes us back to *behaviors* performed within precise contexts (such as buying "too many things" and "substance abuse") that strike the diagnostician as *extreme* (deviant) and *potentially* self-harming. Criteria five through nine include *actions* categorized as suicidal, mood swings that are due to marked *reactivity*, *interior states* (chronic anger and chronic feelings of emptiness) and *illogical* and *inappropriate* beliefs about the self ("transient, stress-related paranoia") and the appearance of *behaviors* that psychodynamic theorists call "dissociated." Among the latter are interior states of mind (a lack of focus, for example) that we ascribe to the patient and behaviors such as "meaningful forgetting," which we cannot easily classify as motivated or unmotivated or motivated unconsciously.

Although I criticize this list of criteria, I do not dismiss it. While it is odd in the extreme, one might say, and subject to revision, it is the best we have at this time. Sophisticated clinicians and researchers can, with proper training, come to use this and similar criteria with genuine skill. In the jargon of the trade, some clinicians can use these categories in ways that are both valid and reliable. Authors of the original *DSM* realized that it was a preliminary effort. After its appearance in 1952, it was revised substantially in 1968 (*DSM-II*) and even more radically in 1980 (*DSM-III*) and again in 1987 (*DSM-III-R*). This 1987 version "utilized data from field trials that the developers claimed validated the system on scientific grounds. Nevertheless, serious questions were raised about its diagnostic reliability, possible misuse, potential for misdiagnosis, and ethics of its use."[2]

Responding to pointed criticisms of its cultural biases and poor integration with neurological theory, *DSM-III-R* gave way to *DSM-IV* in 1994. (*DSM-IV-TR* [text revision] appeared in 2000; it does not alter the diagnostic codes and criteria of *DSM-IV*.) Apart from its cultural biases, learning to use the *DSM* categories is itself a form of acculturation. To employ its categories correctly, one must comprehend the thought world of its authors; this is identical to the task of understanding the rules of a foreign culture embedded in practice and "what everyone knows." Everyone (who has been trained in psychodynamic thought) knows what *illogical* and *inappropriate* mean when describing a person's behavior, the same way that everyone (raised in Protestant, upper-class, white America) knows that the phrase "the honour of your presence" should "be reserved for weddings held in houses of worship, while "the pleasure of your company" is used for weddings in other locations."[3]

The difference between *honour* (using the British spelling) and *pleasure* is, apparently, large; *honour* implies hierarchy and specific location, while *pleasure* implies conviviality and, as it were, egalitarianism. (No doubt I'm missing additional subtleties available to the cognoscenti.) To appreciate the difference between *honour* and *pleasure* and the difference between *appropriate* and *inappropriate*, one must acquire discriminating experiences. On first reading the *DSM-IV*, many a neophyte clinician has exclaimed: "I've done that—sometimes!" or "I've felt that way—sometimes!"

Depending upon what one is told and upon the interpretive schema laid upon one's actions, these diagnoses look distressingly like witchcraft allegations. The authors of the *DSM-IV* section recognize these dangers and take great pains to address the need to sample the patient's behavior in as many complex ways as possible: "The essential feature of Borderline Personality Disorder is a pervasive pattern of instability of interpersonal relationships, self-image, and affects, and marked impulsivity that begins by early adulthood and is present in a variety of contexts." To make this judgment about another person, especially to denote his or her "pervasive pattern" of behaviors, one must know a great deal about that person's life story, especially about his or her forms of love. This cannot be gained in a thirty-minute interview or by a checklist ticked off by a harried nurse. On the contrary, the language of *DSM-IV* is dramatic and novelistic: the clinician searches for patterns just as the literary critic searches for patterns in *Hamlet* or *Little Dorrit*. *DSM* asks us to assess the patient's life story in its most intimate moments: What happened in early adulthood when the patient sought love? How did she or he navigate the tasks of living with others and depending upon them?

This association between diagnosis and pattern recognition sits uneasily amongst scientific types. To continue in the language of the *DSM*: we must discern in the patient's narrative those actions we judge to be reasonable and those we judge to be impulsive. Again, some people can make these judgments, just as some can distinguish great literature from lesser literature, but both tasks require extensive experience and taste. As the *DSM* authors put it, we must assess how the patient operates in a variety of contexts. Not said but implied is that we must assess how the patient's story, assessed in a variety of contexts and through major developmental periods, compares to the normal and expectable story, as we understand the category "normal."

To accomplish that task we must ground our assessment in the warp and weave of a particular life lived in a particular place and time. This makes psychodynamic assessment a cousin to cultural anthropology, and both of them versions of the empathic narrative most often found in great literature.

For example, to ascertain what is normal and not normal in Southern religion in the United States around 1947, we should read Flannery O'Connor's celebrated novella *Wise Blood*. Hazel Motes, a young man fresh from the U.S. Army, comes home to his Southern city and finds himself compelled to imitate the dress, but not the beliefs, of his tent-revivalist grandfather. Enraged by the religious fraud he sees around him, Hazel finds himself founding the "Church Without Christ." In his church "the blind don't see and the lame don't walk and what's dead stays that way."[4] Even as he struggles against church and preachers and Christ, people around Hazel see him as a new prophet, a visionary whom they wish to follow.

Hazel, the antipreacher, the antichurch zealot, and the man who throws himself into a frenzy to deny the reality and power of Christ, is absorbed by the thing he abhors. As the novel progresses, Hazel's path seems bizarre and abnormal. (When he binds himself with barbed wire and blinds himself with lye, he would merit a quick diagnosis of psychotic or a version of delusional disorder, according to the *DSM-IV*.) And yet, while she portrays Hazel as extreme, O'Connor does not portray him as meaningless. At the novel's end, Hazel is accused of not paying his rent (a form of crime according to the law). Two fat cops set out to find him and bring him to justice. They find Hazel, like King Oedipus, blinded by his own hand, lying in a drainage ditch:

> "You reckon he's daid?" the first one asked.
>
> "Ast him," the other said.
>
> "No, he ain't daid. He's moving."
>
> "Maybe he's just unconscious," the fatter one said, taking out his new billy. They watched him for a few seconds. His hand was moving along the edge of the ditch as if it were hunting something to grip. He asked them in a hoarse whisper where he was and if it was day or night.
>
> "It's day," the thinner one said, looking at the sky. "We got to take you back to pay your rent."
>
> "I want to go on where I'm going," the blind man said.
>
> "You got to pay your rent first," the policeman said. "Ever' bit of it!"
>
> The other, perceiving that he was conscious, hit him over the head with his new billy. "We don't want to have no trouble with him," he said. (*Wise Blood*, 230–31)

When Hazel dies at the hand of a cop, we cannot but reflect on the similar death of Jesus at the hands of the Roman authorities. Like Jesus, Hazel's deepest guilt is that he challenged a worldview centered upon civil authority and given infinite power over life and death. Speaking as a critic, O'Connor looked back upon the book and called it comic novel about a Christian *malgré lui* (in spite of himself). She challenged the nontheist reader who would prefer to think that belief in Christ is "a matter of no great consequence." To understand what is normal about Hazel Motes (and about Flannery O'Connor), the critic and the psychiatrist will have to declare themselves upon the issue of Christ's divinity. This is not the kind of task for which medical school prepares one.

If we are retain the category of personality disorders, then we must retain this requirement (which seems to me inescapable): we have to locate our patients within these complex stories and contexts and, at times, take a stance on what Christ means in ways that surgeons and dentists need not. However, to take a stance on the meaning of Christ is to proclaim what is normal, and that kind of proclamation brings us close to the looming cliffs of conventionality. Lacking a set of objective, verifiable symptoms analogous to those of an incontrovertible disease, such as polio, psychiatrists rely upon concepts like *instability* and *irrationality*. While essential to the conduct of civil life, these concepts do not admit of simple or direct validation, much less objective measures associated with science. (While Hazel was guilty of not paying his rent, that fact does not explain his execution.) Psychiatry cannot escape this humanistic challenge. As we shall see, for that reason psychiatrists—our contemporary priests—find themselves engaged in the major religious and ethical battles of our time.

DIAGNOSING THE BORDERLINE PERSONALITY: FIVE OF NINE SYMPTOMS

From the beginning of modern psychiatry, these novelistic and moralistic features of psychiatric nomenclature have roused deep suspicions. In both scholarly and popular accounts of the "history of madness," we confront fundamental problems with medical treatment of insanity (or mental illness, or madness). Even a historian as seasoned as Roy Porter can write a brief history that simplifies the struggle to understand mental illness.[5] More subtle still are historical accounts of the contexts in which madness was assessed and treated.[6] In our times, the most famous and contentious studies of the history of madness and its sociopolitical constructs are Michel Foucault's celebrated books.[7]

Given the vagaries of psychiatric categories and the ways in which they can be used by those in power to label those not in power, diagnoses can appear to be no more than witchcraft allegations. This, in turn, makes those who wield them, and those diagnosed, actors in political struggles. More recently, we recognize the same quality in debates about the propriety of diagnosing homosexuality as a mental disorder. For example, in a recent study on the scientific controversies, three eminent persons discuss the debate over removing the category of homosexuality from the *DSM*: Ronald Bayer, a historian; Robert Spitzer, a distinguished psychiatrist who championed the examination of the issue; and

Irving Bieber, a psychoanalyst and psychiatrist who claimed that his scientific data argued against removing the term from the *DSM*.[8] According to Bieber, this action was "the climax of a sociopolitical struggle involving what were deemed to be the rights of homosexuals."[9] In his lengthy review of the complex reviews and assessments done by various groups of scientists that led up to the American Psychiatric Association proposal, Bayer does not support Bieber's rendition.[10]

With admirable clarity, Spitzer recognizes the mixed language and mixed discourse that make up many of the categories of the *DSM*. He was able to draw upon his prestige as a severe, insider critic of psychiatric diagnoses, established in a series of much-respected studies of the reliability of diagnoses.[11] Even here, though, one can criticize Spitzer's criticism since he and his colleagues labeled kappa, a reliability score, using negative categories, such as "only fair": "Had their purposes been different, Spitzer and Fleiss could have described the state of reliability as quite good. Or they could have emphasized that in every single diagnostic category in every study, psychiatric agreement was considerably better than chance. This observation could have been described as a diagnostic achievement, earlier obscured, but now revealed by the advent of kappa."[12]

Regarding the labeling of homosexuality as disease (or disorder or dysfunction), Spitzer notes that one can support this contention using religious assertions, folk beliefs, and vague speculations about the role of nature in sexuality. In themselves, however, none of these should matter to a body like the American Psychiatric Association, which claims to be among the sciences: "The concept of 'disorder' always involves a value judgment. In the case of homosexuality, it is not possible to reach a consensus at the present time as to whether or not ... [the] inability to function heterosexually because of an exclusive homosexual orientation should be conceptualized as a mental disorder."[13]

The move to depathologize homosexuality had the predictable effect of disturbing a sizable number of Americans of all types and has remained a source of outrage for many on the religious right. Thus we read in a current web page, "Jesus-Is-Savior.Com," that the decision by the American Psychiatric Association to remove homosexuality as a mental disorder from *DSM-II* was a "significant victory for homosexual activists, and they have continued to claim that the APA based their decision on new scientific discoveries that proved that homosexual behavior is normal and should be affirmed in our culture. This is false and part of numerous homosexual urban legends that have infiltrated every aspect of our culture. The removal of homosexuality as a mental disorder has given homosexual activists credibility in the culture, and they have demanded that their sexual behavior be affirmed in society."[14]

Five of nine forms of diagnosis have given increased rigor to psychiatric research and made possible important collaborations, at both national and international levels. For all their utility, though, no one claims that they amount to rigorous science. The five-of-nine checklist indicates quite nakedly that we do not have available a causal theory as to the etiology of borderline personality disorder (BPD). That is, we are not naming an agent that, when present, causes BPD.

On the contrary, as the authors of *DSM-IV* admit, what we call BPD symptoms appear and disappear depending upon the patient's perception of the caregiver. This feature, to which any clinician can attest, is a commonplace in psychiatry (but not in ordinary medicine). For example, a defining characteristic of the BPD is a surplus of anger. A consistent finding of research on BPD persons is that they evoke extraordinary amounts of anger in otherwise calm, professional therapists. Glenn Gabbard writes, "From the earliest days of research on BPD, a clear and consistent diagnostic finding has been that rage is a central problem. Patients will be angry much of the time, and therapists may feel like retaliating in kind, or withdrawing by becoming more aloof or by mentally leaving the room. Treatment will be smoother if the therapist can hold firm and stay in a midposition between withdrawing or retaliating. Much of the countertransference acting out I have seen as a consultant has been with therapists who became so intensely angry at a patient that they could no longer function in a professional, competent manner."[15]

Alan Morgenstern summarizes a long therapy with a borderline woman, whom he names Charlotte, and presents excerpts from her self-observations.[16] Describing Charlotte's family, Morgenstern denotes three generations of massive faults in reality testing and social reality testing. Charlotte's grandmother was diagnosed paranoid psychotic and spent twenty-six years in a psychiatric hospital; her mother attempted to merge with Charlotte and thus punished individuation; Charlotte suffered from severe forms of BPD symptoms. Charlotte describes eloquently a standard feature of the BPD: the demand for complete honesty from the therapist.

While they often feel the need to lie or evade questions and topics that are too dangerous, borderline patients scrutinize their therapists with unrelenting focus and intensity. Any evasions, half-truths, white lies—indeed anything less than complete candor by the therapist—evoke intense anxiety (often invisible to the therapist because of BPD style and the ingrained fear of disclosing vital information). Intense anxiety evokes intense anger. Because the patient's anger may come a day or two or three after the offense given, the therapist struggles to remember, sometimes, the exact sequence of events that led up to the crisis.

This gap between the time of the offense and retaliation for it generates a great deal of countertransference distress and accounts, in part, for the common finding that therapists struggle to not slander their patients. Hence, the label *borderline* often verges on becoming nothing more than a moralizing label, as

Marsha Linehan notes.[17] That the therapist is often angry with the patient, but struggles to deny it, matches the patient's expectations that people deny their real feelings toward the patient.

Describing what she calls "Variety Three" patients (the other two being psychotic and neurotic), Charlotte says: "The physician must be a flesh and blood person. In this case, he may have his foibles and he should preferably have them openly. He may be permitted a bad night's sleep, a cranky wife, a bout of flu. These all constitute a part of the physician's reality, and if he is to furnish a measure of reality to the Variety Three patient bits of trivia are of positive Use."[18] However, if the therapist misleads the patient or is other than completely straight, showing complete candor and honesty, all hell breaks lose. Charlotte calls all forms of evasion playing games. If the patient catches the therapist in such a game, "the patient will be merciless, and if the game has been particularly noxious the patient may seek to destroy the doctor psychically."[19] She adds immediately, "I should stress at this point that no physician should embark on treating a Variety Three patient unless he has the quality of candor. This I think takes enormous courage."

The *DSM-IV* authors offer a causal account of borderline anger, saying that it "is often elicited when a caregiver or lover is seen as neglectful, withholding, uncaring, or abandoning." Again, while my clinical experience matches this generalization, I note that it is couched as both a claim of causality—perception of abandonment threats elicit anger—and yet not; "often elicited" means that this account is only a guideline. The *DSM-IV* dictum states, effectively, whenever a patient you believe merits the diagnosis "borderline" flies into rage, see if she or he probably believes that you are threatening to abandon her or him. This dictum is a useful heuristic device—a generalized way to consider what we should look for and how we might intervene with a patient who is busy screaming at us.

This focus on belief (a vastly complex phenomenon) returns in the concluding comments about borderline rage. Anger and other forms of distress may disappear: "The real or perceived return of the caregiver's nurturance may result in a remission of symptoms." Now since the presence of symptoms (at least five of nine) defines the alleged disease, their absence is more than a remission. For the concept of *remission* derives from physical medicine; it means that the underlying disease process, for example cancer, is inactive at the moment. It does not mean cure; the disease lurks within, and until one can say it has disappeared, patient and caretaker continue to worry. When employed in the *DSM* context, *remission* is a metaphor drawn from medicine. By using it the *DSM-IV* authors announce their beliefs that something analogous to cancer or bacteria or some other entity causes borderline symptoms. This hidden cause continues through those periods when the patient is not yelling at us, when his or her symptoms are in remission. While this metaphor is quite clear, its validity as a model of the disease called character pathology is not.

To the dismay of everyone who ever attempted to find rational order in the study of behaviors, ranging from group behaviors to the actions of persons designated as "pathological," bringing rigorous objectivity to these endeavors is an immense challenge. As I suggest throughout this study, we cannot answer that challenge because the objects of our investigation, these diverse and complex human actions, are conditioned by numerous variables. They are not structured like the objects of nature. We cannot investigate them with traditional scientific rigor because we cannot begin with a common set of agreements as to *what* we're measuring. Indeed, many disputes in psychiatry, for example, turn on fundamental questions about the validity of *all* of its categories. For at least fifty years, distinguished critics of *DSM* categories have challenged its dependence upon this disease model of psychological suffering. Among these critics are Erving Goffman, Thomas Szasz, and Thomas Scheff.[20] They have not let up. In a recent review of the claims made for *DSM-IV*, Thomas Scheff summarizes three assertions that, he says, are both affirmed in the media and by the psychiatric establishment and happen to be false:[21]

1. The causes of mental illness are mainly biological.
2. Types of mental illness can be coherently classified (*DSM-IV*).
3. Mental illness can be treated effectively and safely with psychoactive drugs.

Writing thirty-three years after his book first appeared, Scheff does not give reason to believe that psychiatric nomenclature and the search for a causal theory of mental suffering have entered into a scientific golden age. The complex, murky, and confounding problems of diagnosis remain.

ON THE PLEASURES OF SCIENCE FICTION: JUMPING INTO *THE ABYSS*

Controversies within psychiatry persist not because psychiatrists and psychologists are, as a group, dumber than other scientists; nor because there is as vast conspiracy to deny truths such as "all we need is love" and replace them with overpriced pharmaceuticals. There is venality in some of the members of the American Psychiatric Association, but not more than among some members of the American Philosophic Association or the United Methodists. The history of psychiatry and its cousins reveals the usual mix of genius, failure, and confusion that marks any human enterprise. It also reveals numerous efforts to name the cause of psychological disease, what earlier centuries called madness. No such problem appears when we investigate the seven-of-nine phenomenon. United Paramount Network can avoid these messy controversies because it luxuriates in the world of imaginary beings.

Consider *The Abyss*, a science fiction film written and directed by James Cameron in 1989. In it we find features common to other mythlike artifacts: it conveys core conflicts of its native culture (the United States in the late 1980s), it recognizes human limitations and irrationalities, namely the danger of nuclear war, and it offers a fantastic solution to those difficulties.

My account does not capture the viewer's experience of claustrophobia and terror. A huge amount of money spent on special effects and underwater photography makes possible a high level of verisimilitude: you feel yourself trapped in a nuclear submarine, then in a smaller underwater vessel, then in an even smaller vessel as it plummets into an abyss far below the ocean's surface. The notion of "abyss" is embedded in Western religion. From the Latin *abyssus* and Greek *abyssos* ("without a bottom"), it evokes two diametrically opposite valuations. In the first it is an analogue of the human mind in its darker aspects, a point the film draws upon but does not belabor. A fear common to some patients new to psychotherapy is that by entering it they might fall into a deep, dark, place—an abyss or a "snake pit"—from which they will never return.[22] In the second evaluation, *abyss* refers to the "Deep," to *tehom*, the original waters named in Genesis 1:2: "And the earth was without form, and void; and darkness was upon the face of the deep. And the Spirit [or breath] of God moved upon the face of the waters." Hence, in the Latin version of Psalm 42 appears the phrase *abyssus abyssum vocat*, but resonance and mutuality replace anxiety:

> Deep calls to deep at the thunder of your cataracts;
> all your waves and your billows have gone over me. (42:7)

Through obvious associative linkages, "abyss" can also refer to the grave. The deep is both a source of anguish and terror and, yet, when confronted by the power of God, subdued.[23] Like all other ancient peoples, the authors of the Psalms developed cosmologies about the origins of the waters that surrounded them. While the Hebrew creation stories share elements of these stories, theologically they differ. In the biblical text, God rules all without challenge, even the monsters of the Deep. In the Sumerian, Akkadian, Ugaritic, Phoenician, and Greek accounts of cosmogenesis are violent battles, deities contending with deities, not the singularity of God's power. It is in *this latter sense that abyss appears throughout the Psalms*: "Thou, which hast shewed me great and sore troubles, shalt quicken me again, and shalt bring me up again from the depths of the earth" (71:20); "The waters saw thee, O God, the waters saw thee; they were afraid: the depths also were troubled" (77:16); and "Whatsoever the LORD pleased, that did he in heaven, and in earth, in the seas, and all deep places" (135:6).[24]

The Abyss portrays conflicts dominant in Reagan-era America, the late 1980s. The U.S. and Soviet Cold War is still hot, and the U.S. and USSR navies harass each other. A submarine chase off the coast of Cuba ensues. Under tense battle conditions an American submarine, the USS *Montana*, pursues what its captain assumes is a Russian submarine, but then learns that the entity is traveling at 130 knots underwater, a speed impossible for any terrestrial ship. The *Montana* crashes, drowning its crew. This claustrophobic scene announces a constant theme of the film: death by drowning and the wish to transcend it. Nature enters the picture in the form of a hurricane named Frederick. The storm drives the action since it makes possible the introduction of the heroes (civilians) and the villains (U.S. military types). The good guys include a white engineer and a black underwater-craft operator who run an oil rig. The bad guys are Navy SEAL divers sent to cajole the civilians into helping them rescue the submarine's crew and surreptitiously to retrieve the *Montana's* nuclear warheads and to investigate the alien craft.

In usual mythic fashion, the good people share one set of characteristics; the bad guys share the opposing ones. For example, here are some of the film's dualisms:

Good Guys

virile but kind to women
independent working men and women
easygoing, witty, sexy
wear blue jeans with gloves in back pocket and Yankee baseball cap
love and play with pet rat
racially and sexually mixed
distrust nuclear weapons
open—apparently "crazy" but secretly sane
generous to new ideas and new forms of being
love country music
bound to each other as human beings
suspicious of U.S. government
wish to protect new forms of life, the nonterrestrial intelligence

Bad Guys

macho and cruel to women
Navy SEALs, government types, storm trooper mentality
deadly serious and morose
wear uniforms with knives in pants and leather holsters and dark SEAL hats
hate animals; appear to drown the pet rat

all white men
love and wish to use nuclear weapons
closed—apparently rational, but secretly insane, "buggo"
taciturn, duty bound, rigid, hate and distrust new forms of being
love no music; polish weapons
bound to the "mission," not to any person
U.S. military paranoia about "others"
wish to destroy what is new, the nonterrestrial intelligence

The film presents multiple conflicts: storm vs. humans; Russians vs. Americans; SEALS vs. working men and women; alien nonterrestrial intelligence creatures vs. humans; and the husband and wife alienated from one another. Given these conflicts and the mythic dualisms around which the action is structured, it follows that the alien beings will unite in themselves characteristics that Aristotelian (or terrestrial) logic declares incompatible. When the heroine, Lindsey, first touches the aliens' craft she says it was both a machine and yet alive. As the antithesis of Coffee, the cruel warrior, Lindsey reveals her bravery when she tastes one of the alien "water creatures" after it invades their underwater capsule. Seeing the same creature, Coffee panics and attacks it.

To resolve each of these conflicts the dramatis personae must be born again in a literal sense. Like other mythic heroes, they must do the impossible—that is, make impossible choices and pass beyond impassable barriers. Prefiguring the theme of death, descent, and resurrection, we see the white rat subjected to what all mammals feels is certain death: it is forced to breathe a liquid. One Navy SEAL, the lesser of the bad guys, explains that the liquid is highly oxygenated and that "you breathed liquid for nine months—you never forget how."

As William James observed in *The Varieties of Religious Experience* (1902), religious quests typically demand that the one seeking transcendence die and then be born again, just as Jesus died and was born again. ("Except a man be born again, he cannot see the kingdom of God"; John 3:3.) In *The Abyss* this cliché is revitalized when first the woman, then the man, must drown to save others. Realizing that both she and Bud cannot swim back to the underwater vessel—having dispatched Coffee to his grave in the abyss—Lindsey volunteers to drown, to go unconscious in the extremely cold water and thereby conserve oxygen. Then, using the remaining oxygen, Bud can tow her underwater and revive her once they are back on the diving platform. To save them both she risks everything. This theme reappears later when, in order to save the alien beings whom they learn are at the bottom of the abyss, Bud volunteers to swim down and disarm the nuclear device Coffee has dispatched there. Bud realizes that he will have too little oxygen to return from his mission and that he will certainly drown. We see him accomplish his mission and then faint. Trying to duplicate his rescue of her, Lindsey shouts at Bud, "You dragged me back from a bottomless pit," that is, the deathlike state she entered when he towed her underwater.

Finally, the themes of sacrifice, true love, and resurrection unite in a transcendental moment when the alien vessel ascends from the abyss, saving the hero and the entire diving crew. Not accidentally these themes, especially the ultimate reliance upon transcendent beings to resurrect the living and the dead, are staples of traditional religious promises. Bud's Christian name, as one says, is "Virgil"—a reference to his pristine character. The alien beings who rescue him are winged creatures that resemble medieval portraits of angels.

The same theme of human and extrahuman, or human and seemingly human, appears throughout the *Star Trek* series. *Voyager's* frequent shoot-'em-up battles and sexual scenes are subordinate to Roddenberry's fascination with the extension of our notions of human. An easy way to do this is to pair a normal (though brave) human with a not-fully-human counterpart. Thus, in the first television series, *Star Trek*, Captain James T. Kirk, a robust tough guy given to emotionality, is paired with an austere, hyperrational Commander Spock, a half-human, half-Vulcan: "Spock's mother, Amanda Grayson, was a human schoolteacher from Earth and his father, Sarek, was a respected diplomat."[25]

At its fan site, Paramount provides detailed notes on Seven's background, her parents, and her role on the Voyager space ship (technically, a Starship, according to the mythology of Star Trekdom).[26] Seven is a wildly popular character, given the attractiveness of Jeri Ryan and the sprayed-on costume she wears. Indeed, while her sexual allure is essential to the character, Seven became more interesting when her creators made her part human, part cybernetic.

In the nineteenth century Seven would have been part human and part animal; or part human and part electrical device; or part human and part of another human. Whatever conflictual pair that grips a culture at a given time about the nature of human nature reappears in its folk fiction and its science fiction. (We can expect, therefore, a science fiction character who switches from male to female or vice versa.) When the idea of human cloning became a topic of public speculation, Hollywood followed immediately with a couple dozen films. These range in quality, from the competent like *Star Man* (1984, directed by John Carpenter) and *Jurassic Park* (1993, Steven Spielberg), to the bizarre like *The Clones of Bruce Lee* (1977, Joseph Velasco), to the comic like *Multiplicity* (1996, Harold Ramis). Each is an effort to assimilate the new scientific possibility to our commonsense view of the world and the notion of human being.[27]

Although Seven is not a clone, she is that other possibility, a human being who has become a machine, dominated by sentient machines. As the official myth/record put it, Seven was a normal human child, named Annika Hansen, until she and her parents were "assimilated" into the Borg. The Borg is a relentless machine-organism, a collective that spreads like cancer through the universe, sucking all sentient species into its single consciousness. It even assimilated (temporarily) the heroic Jean-Luc Picard, Captain of the *USS Enterprise*. To keep those it assimilates enslaved, the Borg implants an "upper-spinal column neurotransceiver" into them. Through this and other devices, the Borg controls the subject, overriding his or her sense of independence. What science fiction imagines and shows, contemporary authors ponder as just around the corner. Some, like Ramez Naam, champion the idea of "transhumanism" as new improved form of human being; others, like Francis Fukuyama, warn us in apocalyptic terms that this is precisely a nightmare that will soon confront us.[28] In the Trek universe, the nightmare has come true, and viewers experience it through Picard's assimilation and eventual liberation.

Characters like Jean-Luc Picard recognize within themselves mixed qualities of heroism and cowardice. When the brave captain joins the Borg we know that it is against his will; he will eventually conquer the darkness and rescue the others (though not before killing a few of the good guys). We, his fans, would have it no other way. In his actions as part of the Borg, Picard resembles Shakespeare's Richard III, a flawed hero who fascinates us even as he seduces the wife of a man he has murdered. Other science fiction narratives commonly regress to cowboy and Indian narratives. This ploy, this need to keep heroes separated from villains, moves science fiction away from literature and away from history. (Although George Lucas attempts to make his villain, Darth Vader, more understandable and thus interesting in the fifth and sixth films in the *Star Wars* saga.)

While five of nine forms of diagnoses are partial and can lead us awry, they are efforts to comprehend the inner duality of human lives. *DSM-IV* diagnoses are not complete; they do not easily rise to the level of science; they often merit the severe criticism heaped upon them. Yet, they strike me as valuable failures. We cannot say the same about typical science fiction. These accounts and images are satisfying—as escape and as daydream—but ultimately tell us nothing new about human beings. By disowning parts of the self that evoke disdain, some science fiction narratives preserve a split in the self. This serves an obvious defense; it is not I who feel these aggressive or sexual wishes, it is the Other, the Stranger who is enough unlike me to warrant projecting my despised self upon him or her. Picard's merger with the Borg, like Seven's mixed parentage, transcends this defense by showing us characters identified with the Other.

Thanks to Freud and others, we recognize that this deep wish to disown parts of the self and to project those parts onto others drives both individual and group pathologies. Among the former is the borderline personality disorder; among the latter are numerous forms of racism and hatred of the Other. The dreadful brilliance of *Mein Kampf* is that Hitler merged himself with the idea of health and Jews with the idea of disease.[29] Readers who grant the premises of *Mein Kampf* and who share the humiliation Hitler evokes in that text find themselves in a Looking Glass World. It is topsy-turvy not just because ordinary values are replaced with totalitarian ones, but because the divisions within the self are healed, and distances between persons is collapsed. Arthur Feiner described the new form of oppression created in the Nazi state:

> Not only were the new technological means of total domination over people more drastic than ever before, but this totalitarianism differed from other forms of political oppression (despotism, tyranny, and dictatorship) in that "Wherever it rose to power, it developed entirely new political institutions and destroyed all social, legal and political traditions. . . ." (Arendt, 1958). Arendt points out that under the Nazis, the body politic, the place of positive laws was taken by total terror, which made it possible for the so-called force of "nature" or "history" to run freely through a society unhindered by spontaneous action and response. Thus the space between people disappeared, and humans were pressed together (after they had been isolated from each other) to become One.[30]

Once bonded to each other in this way, a past is made and escape is not easy.

Figure 4.3. Oedipus and the Sphinx.

PROGRESS AS DEVELOPMENT OF THE SELF:
FROM GREEK CULT TO GREEK THEATER

It is not accurate to say that everyone who indulges in splitting is on his or her way to joining the Nazi party. Counting myself among the millions who enjoy science fiction, I do not find myself edging toward fascism. Nor is it fair to say that science fiction is inherently fascistic. However, we can say that part of Hitler's power derived from his ability to exploit the pleasurable relief that splitting and projection afford. The genuine pleasure that splitting offers is relief from the awfulness of internal conflicts—and the struggle to understand our minds. In this way, science fiction offers an escape from interiority. We will explore strange new external worlds, populated with bizarre, humanlike entities who live out there, not in here. We share the front seat with the omniscient narrator who flies past ordinary limits and conjurers up faster-than-light spacecraft, sword fights, and epic battles in which we join the valorous, strong, and unfairly maligned virtuous. As in any good story, the heroes must suffer many, nearly calamitous defeats before they triumph in the last hour of the last battle. Like the heroes of *The Abyss*, we must pass through what seems certain death, in the depths, before we can find rescue from transcendental forces. Given the moralizing dualities of science fiction and myth, we identify with the forces of good. Like Agamemnon and Luke Skywalker, we seek justice not revenge.

For all the many pleasures this gives us, splitting makes learning impossible. Having denied and split off what is in us, we cannot find in those projected contents the truth of their origins. To learn more about our interiority, we turn to the arts. We have seen one instance above, in Shakespeare's portrait of Richard III. Similar portraits occur in Greek epic and Greek tragedy; similar struggles appear in the Modernist interpretation of Greek drama as represented by Freud's recasting of the Oedipus story.

I focus on the Greeks because of my training in institutions that idealize the Greeks and because of my location in an American university. The patterns I ascribe to the Greeks may well appear in other traditions and in other literatures. Indeed, if these patterns are valid and not merely projected onto the Greeks, they *should* appear in other literary and religious traditions. Thanks to the rise of feminist theory and other forms of reflection, we note that splitting also occurs regularly in scholarship and the canon of the West. I was taught that the Greeks (of the classical period) gave us democracy, science, and most importantly philosophy—as opposed to the Orientals, the barbarians against whom they fought. In the language of splitting, the Greeks were good and noble; the barbarians were bad and destructive. The ways in which Europeans and their offspring have idealized the Greeks is a long, complex, ongoing story.

That said, we can find something to learn by comparing Greek epics and Greek drama against one another and both against Freud's reading. For in all three genres we find versions of splitting and consequently dualisms without end. Feelings of being overtaken by an Other (a Borg-like force), or being split into opposing characters, are remarkably similar to accounts of possession in Homeric religion. There we find a conception that all deeply emotional states of mind, especially sexual mania and rage in which one acts "against reason," are the work of the gods. In a fine moment of Modernist hope, we might seek to find a line of development that stretches from Greek religion, to Greek drama, to Greek philosophy. This was the task that Bruno Snell, a German classicist, assumed in *The Discovery of the Mind: The Greek Origins of European Thought.*[31]

Snell claims to delineate a progressive, evolutionary track that began with Greek mystery religions, which were cultic and depended upon deliberately induced dissociative experience, and eventuated in Greek tragedy. In Greek tragedy, he argues, we find that the central goal, the driving ambition, is to achieve a single, lucid consciousness. This unified self exemplifies the Western ideal of the person who has left religion, especially cultic and ritual practices, behind. For cultic celebrations reduce and sometimes dissolve consciousness; their allure is that they grant us reprieve from the search for self-knowledge and unity.

Reflecting upon Greek literature, Snell claims to show that this transformation occurred between the Homeric period, about 1000 B.C.E., and the pinnacle of tragic theater, about 400 B.C.E. How did the Greeks evolve their introspective masterpieces, including Sophocles' *Oedipus Tyrannus*, which dominate Western thought, from what were once mystery rituals? "In Homer a man is unaware of the fact that he may think or act spontaneously, of his own volition and spirit. Whatever 'strikes' him, whatever 'thought comes' to him, is given from without, and if no visible external stimulus has affected him, he thinks that a god has stood by his side and given him counsel. . . . In the tragedies of Aeschylus, on the other hand, the agent, conscious of his individual freedom of choice, makes himself personally answerable for his actions."[32]

Being personally answerable does not mean the hero claims complete responsibility for his actions. To heirs of the Greeks, the Renaissance and Freud, these two forms of accountability should be the same thing. We require agents to discover their wishes and thus to discover their role in their actions; Homer does not: "When the Homeric hero, after duly weighing his alternatives, comes to a final decision, he feels that his course is shaped by the gods."[33] From our Modernist perspective, this is unacceptable because the gods are not external beings who live a long distance away from us, on a mountain. For the Homeric hero, speaking from within his experiences, to not believe in the gods is equivalent to not believing in the most vivid part of his experience. Only a fool or mad person would deny this reality. Greek gods are neither magical nor unfathomable. Like humans, they are motivated by passions, especially envy and the lust for revenge; they are subject to the work of fate, as well. They do not create ex nihilo: "The supernatural in Homer behaves with the greatest regularity."[34]

Because he is typically disconnected from the emotional source of his actions, the Homeric hero does not feel the weight of guilt for his actions, especially those carried out while possessed by a *daemon* or other nonself entity. In contrast, some six hundred years later, tragic heroes gain stature to the degree they feel connected to the consequences of their actions. However, even then being connected to one's fate, recognizing the working out of bloodguilt, for example, does not mean that one feels responsible and thus guilty. This becomes evident in *Oedipus at Colonus*.[35] Following the discovery of his great errors chronicled in *Oedipus the King*, Oedipus was cast out of the city. In *Oedipus at Colonus*, the aged Oedipus addresses those, especially King Creon, who accuse him of incestuous guilt and parricide:

> *I have suffered it all, and all against my will!*
> *Such was the pleasure of the gods, raging,*
> *perhaps, against our race from ages past.*
> *But as for me alone—*
> *say my unwilling crimes against myself*
> *and against my own were payment from the gods*
> *for something criminal deep inside me . . . no, look hard,*
> *you'll find no guilt to accuse me of—I am innocent.*[36]

Oedipus can make this claim because like the gods and other human beings, he shares passions, memory, and sometimes unknowable wishes. To justify killing Laius, his father, who charged at him in a silly roadside altercation, Oedipus asks the Chorus, "If here and now, a man strode up to kill you, you, you self-righteous—what would you do? Investigate whether the murderer were your father or deal with him straight off? Well, I know, as you love your life, you'd pay the killer back, not hunt around for justification" (1132–38). The instinct of self-preservation and immediate action, without which one could not be a warrior and a *Tyrannus*, requires self-defense. This justifies Oedipus's warlike action, and while it had the unintended consequence of killing his biological father, it does not merit blame according to the heroic code.

In the earlier play, Oedipus rages at Tiresias, who refuses to tell him his secret knowledge of the meaning of the plague that afflicted the city and of Oedipus's fate. His flaring temper that had incited him to kill Laius provokes Tiresias and thus furthers the horrible fate that awaits him: "You [Oedipus] criticize my temper . . . unaware of the one you live with, you revile me" (*Oedipus the King* 384–85). To Oedipus's plea to learn the future, Tiresias replies: "I will say no more. Do as you like, build your anger to whatever pitch you please, rage your worst—" (391–92).

When Oedipus killed Laius at the crossroads it was in response to being insulted by Laius's driver. Oedipus responded to the "brute force—and the one shouldering me aside, the driver I strike him in anger!—and the old man, watching me, coming up along his wheels—he brings down his prod, two prongs straight at my head! I paid him back with interest! . . . I killed them all— every mother's son!" (888–97). Oedipus's blinding rage, his self-love and pride, causes him to return mild aggression with violence and reprimand with utter terror; because he felt insulted he killed four men.

The Chorus calls this "Pride" (*hybris*). In one of the most famous lines in all of Sophocles they say: "Pride breeds the tyrant, violent pride gorging, cramming to bursting with all that is overripe and rich with ruin, clawing up to the heights, headlong pride crashes down the abyss—sheer doom" (963–67). Just prior to this famous line appears the larger theological claim about Destiny (*Moira*), to whom the chorus pleas:

> *Destiny guide me always,*
> *Destiny find me filled with reverence*
> *pure in word and deed.*
> *Great laws tower above us, reared on high*
> *born for the brilliant vault of heaven—*
> *Olympian Sky their only father,*
> *nothing mortal, no man gave them birth,*
> *their memory deathless, never lost in sleep:*
> *within them lives a mighty god, the god does not*
> *grow old.*[37]

Snell reaffirms the Western adoration of the late Greeks when he says tragic theater developed beyond the cultic attack on ego unity and its championing of a dissociative aesthetic. For *Oedipus at Colonus* is crucial to Sophocles' theology; it articulates the nature of self-consciousness and the cost of that new kind of knowledge. Oedipus's achievement is typically Greek and its cost, banishment, fits the crime. Having borne the suffering that wisdom requires, Oedipus returns to favor after many years and achieves a final apotheosis. In final obedience to the "great laws that tower above us," Oedipus finds rest.

He does not have to plunge beyond that theological truth. This religious answer is not available to Modernists. We see this difference in the earlier plays; for example, when attempting to quiet Oedipus's fears prior to the awful revelation, Jocasta answers the terrible claim that Oedipus has actually shared his mother's bed: "And as for this marriage with your mother—have no fear. Many a man before you, in his dreams, has shared his mother's bed. Take such things for shadows, nothing at all—" (1073–76). Here, Modernism says, using Freud's voice, is Jocasta's error. She denies to dreams any possible linkage with action—a belief opposite to the usual Greek ascription of occult power to dreams—and she denies that more than chance rules our lives. If true, this claim would contradict the theology of the plays: that Oedipus was destined to sin because of his actual blood linkage to the past, the curse on his house, and it denies that the oracles could have known the future.

If valid, Jocasta's admonition would also deny the core theorem of psychoanalysis, that there is a meaning to dreams and to other partly formed expressions, emissaries from the preverbal heart of human being. As children of Freud, Modernists cannot affirm oracular religion. Indeed, we deny both the premise and the conclusion of the play. We deny that there was a prophecy from on high; we deny that Moira caused Oedipus to carry out his terrible deeds. Rather, with Freud, we assert it was an insight from below. We cannot fully agree with Snell: "Tragedy is not a faithful mirror of the incidents of myth; instead of accepting them as historical reality, as they are in the epic, tragedy traces the ultimate causes in the actions of men [sic], and consequently often pays but little attention to concrete facts."[38] But this is precisely what is at stake: "The ultimate causes in the actions of men."

As every critic of the plays has noted, Oedipus is the tragic hero who seeks knowledge above all else.[39] The tremendous condensation of Sophocles' poetry and the pell-mell plot admit of numerous, persuasive readings of Oedipus's actions. We see readily that attacking the eyes is, typically, a displaced reference to castration and that if Oedipus had castrated himself he would have cut off the offending organ with which he committed his second great crime. Yet attacking his eyes is equally about an attack upon the center of knowledge, his pride in "sight" and "foresight." We recall that Prometheus, the great demigod and culture hero who brought foresight to human beings, was also punished for this transgression (Prometheus means "foresight"). Vision for the Greek was an active, assertive (phallic) action of sending out "rays" from one's eyes upon the world. In contrast to our modern story, taught since childhood, that light strikes us and our "passive" retinas, Greek physiology held that looking and seeing were active, agential practices.

Having solved the mystery of the sphinx and given the Greek answer "human," Oedipus braves a battle with deeper wisdom represented by Tiresias, himself blind and a sexual outlaw, having been both male and female. Like Seven of Nine, Tiresias manifests in himself a duality that seems permanent. Oedipus's greatness was to persist in searching for the truth, for the guilty party. He pursued the great mystery to the end: why was his city cursed? In more theological language: why are we humans cursed? Why do we feel obliged to honor the ancient Greek proverb "better to have never been born"? In Tiresias, the blind, double-sexed seer, Sophocles locates an avenue to transcendent knowledge and thus represents traditional Greek religion. As Oedipus addresses him: "You, my lord, are the one shield, the one savior we can find" (*Oedipus the King* 344–45). As Snell says, the cost of this greatness and this increasing discernment is loneliness: "The deity speaks from a greater distance; man begins to ponder the mystery of the divine, the more isolated he becomes. The characters of Sophocles' plays are already lonelier than those of Aeschylus."[40]

Although Oedipus is lonelier, perhaps, than the archaic characters in Aeschylus, he is not bereft. By doing his duty and by protesting, correctly, his innocence, Oedipus undoes his banishment. He has paid the cost and he merits reentry into the city, into the Polis. In this sense Oedipus represents the height of Greek virtues, at least Athenian virtues, as denoted in Thucydides' *History of the Peloponnesian War*, which he began writing when the war began in 431 B.C.E. and up through its end in 404 B.C.E. Believing that "it would be a great war and more worthy of relation than any that had preceded it,"[41] Thucydides recounts Pericles' Funeral Oration for the Athenian Dead given around 430, a year after the war began. We can read this canonical text any number of ways. Garry Wills, for example, notes that the oration prefigured numerous speeches that drew upon it, including Lincoln's Gettysburg Address delivered some twenty-three hundred years later.[42]

If we ask of this text what chief virtues it celebrates and how these pertain to the unity of the self, we discover that the self is unified only when it is bonded to others of an idealized past. We learn, says Pericles, that the Athenians are superior to barbarians and other Greeks because they recognize both pain and pleasure, but do not let the first expunge the second. Lest anyone believe that he has chosen this occasion for selfish or prideful reasons, Pericles says that the task of the funeral oration was given to him by the ancestors: "Most of those who have spoken here before me have commended the lawgiver who added this oration to our other funeral customs. It seemed to them a worthy thing that such an honor should be given at their burial to the dead who have fallen on the field of battle."

Warning the listener that his eloquence may be inadequate to the greatness of the dead, Pericles notes that he might say too little or too much: "The friend of the dead who knows the facts is likely to think that the words of the speaker fall short of his knowledge and of his wishes; another who is not so well informed, when he hears of anything which surpasses his own powers, will be envious and will suspect exaggeration." Yet, since custom requires it, he will speak, but first of the ancestors: "For it is right and seemly that now, when we are lamenting the dead, a tribute should be paid to their memory. There has never been a time when they did not inhabit this land, which by their valor they will have handed down from generation to generation, and we have received from them a free state."

They and the generation immediately prior improved Athens and made it great by following certain ethical and judicial rules: one is democracy, another is reverence for tradition and unwritten laws of conduct that guide proper action: "While we are thus unconstrained in our private business, a spirit of reverence pervades our public acts; we are prevented from doing wrong by respect for the authorities and for the laws." More so, while other Greeks states, most notably Sparta, enforce laborious and endless effort preparing for war, Athens does not: "Our military training is in many respects superior to that of our adversaries. Our city is thrown open to the world, though, and we never expel a foreigner and prevent him from seeing or learning anything of which the secret if revealed to an enemy might profit him. We rely not upon management or trickery, but upon our own hearts and hands."

This form of openness and trust appears in the Athenian love of beauty; it makes them stronger, not weaker: "If then we prefer to meet danger with a light heart but without laborious training, and with a courage which is gained by habit and not enforced by law, are we not greatly the better for it? Since we do not anticipate the pain, although, when the hour comes, we can be as brave as those who never allow themselves to rest; thus our city is equally admirable in peace and in war. For we are lovers of the beautiful in our tastes and our strength lies, in our opinion, not in deliberation and discussion, but that knowledge which is gained by discussion preparatory to action. For we have a peculiar power of thinking before we act, and of acting, too, whereas other men are courageous from ignorance but hesitate upon reflection."

The greatest danger is giving into fear and acting in a cowardly way—merely to preserve one's life and in so doing sever the bonds that unite the self with the ancestors and Athens's greatness. Far worse than death is the succumbing to mere instincts, which damages the person and diminishes the protection offered by union with the city's greatness: "I believe that a death such as theirs has been the true measure of a man's worth; it may be the first revelation of his virtues, but is at any rate their final seal. For even those who come short in other ways may justly plead the valor with which they have fought for their country; they have blotted out the evil with the good, and have benefited the state more by their public services than they have injured her by their private actions."

By preserving these civic virtues, the honored dead live with us, in our greater spirit: "The sacrifice which they collectively made was individually repaid to them; for they received again each one for himself a praise which grows not old, and the noblest of all tombs." According to the rhetoric of the speech, then, the honored dead are not obliterated: they persist as long as we and our memory persist. Here, one might say, the paradox of the immortal self persists: we gain immortality by dying. If we can follow out the heroic code of the city, including the drive toward pleasure, we will win our place in the deathless narration. Here civic duties replace cultic duties.

While Freud focused upon unconscious sexual wishes, his greater task was to challenge both Greek religion and Greek self-adoration. Snell's adoration of the Greeks parallels that of Goethe and Nietzsche, but Freud brings something new to the discussion. In rereading Sophocles as an adult, Freud eschewed his boyhood hero worship. He focused upon sexual knowledge and sexual wishes that Oedipus exemplifies in his actions, even if not his conscious beliefs, in *Oedipus the King*. Freud rejects Jocasta's admonition that Oedipus ignore his anxiety about incest dreams, and he rejects the dualism of Greek religion exemplified in Sophocles. The singular force of *The Interpretation of Dreams* (1900) is Freud's insistence that through the labor of psychoanalysis we can retrieve our part in the production of dreams and, unlike Oedipus, our part in incestuous fantasy. For as strong feelings were to the archaic Greek, so too dreams are typically understood in folk psychology to be "from the beyond." Their messages and contents, their occult powers, derive from the phenomenological fact that during sleep dreams appear to be given to us the way that Oedipus describes his unfortunate life.

If one agreed with this almost universal folk belief, the rational interpretation of dreams would be impossible. We would find ourselves back with the Greeks: either dreams come from entities beyond our mind or they are meaningless. Affirming either claim makes interpretation almost impossible since there is no obvious and evident way for the person to reconnect the sequence of experience, pleasure, and terror that led up to the dreaming experience. The oracular engines of the Oedipus legend are the prophecies that drove Laius and Jocasta to reject their infant son. Without faith in the power of these prophecies the story could not have begun nor could the plot progress, since when Oedipus hears the same prophecy it drives him from his (foster) parents' home, headlong toward Thebes and his meeting with Laius on the road. This central aspect of the play is its religious core. The work of fate, or *Tychē*, is beyond the reach of humans and gods.

Like dreamers, suffering persons cannot know their own minds. If empathically attuned and observant, another person can help us focus upon these moments of pain, retreat, and defense that eventuate in dreams and symptoms. Over time we can discover those parts of our experience we have defensively "repressed" or "denied" or "split off." Implicit in Freud's work is the affront to the naive self-consciousness that follows a more or less successful defense. From within, defensive maneuvers may feel seamless, and thus the gap that had occurred only moments before is invisible to the naive, self-observing ego.

Freud's empathic discoveries included his observations that when we deny a part of our mental life another, unconscious part of the mind registers this as an attack upon the integrity of oneself. Repression and all other forms of defense against insight leave tracks upon the ego's functioning.

Discovering that track, in opposition to the patient's wishes and self-understanding, is a worthy goal because it promises an eventual unity within the self. At the same time we seek the unity of the self we seek unity between self and nature. But in searching for this unity, for a singular truth announced in a clear and unambiguous language, we rediscover our dividedness, our animal nature, our deep aversion to reason itself. Honest observers, according to the Greeks and to those shaped by Greek thought, must confess that their minds are bifurcated; one part seeks unity and wisdom, the other part actively resists unification. In the ongoing conversation about self (or ego) and its struggles that began at least with Plato, Western thinkers struggle to reconcile these oppositions within the self and between self and nature. Typically the latter appears as the struggle between reason and nature.

But that is precisely the issue at hand for Freud and for Modernism. If we agree that human beings have internal agencies, ranging from secret wishes to what Freud called the "id," then by expanding the scope of what counts as part of human being psychoanalysis expands the scope of responsibility. Thus in the "Rat Man" case history (1909), Freud notes that the central cause of his patient's obsessional symptoms lay on the surface of the patient's narration. Because of his patient's disorganization and the effects of these massive, repressive regimes, Rat Man could not see that these superficial elements in his autobiography were the cause of his suffering. To cure him Freud had to expand his patient's understanding of personal causality beyond the boundaries established in *Oedipus at Colonus*.

In the language of classical Freudian theory, repression and other forms of archaic ego defense impede the ego's tendency toward integration and its opposition to dissociation. Accurate interpretations of defenses, like repression and denial, can be empathic because they reflect the therapist's ability to side with the part of the patient's mind, what Freud termed the observing ego. We are aware, sometimes, of having been harmed even if one cannot say precisely how that harm took place. It is not empathic to agree simply with the patient's defensive maneuvers, to join the patient in denial, for example, because to do so denigrates this part of the patient's deepest self-experience.[43]

The accurate and well-timed interpretation of a resistance, such as denial, is relieving because it is empathic. It is also frightening, first because it evokes the original pain and because it reveals a new, not yet mastered demand. If the gods are dead or if dissociation no longer is available to the abused patient, then a new, real, and external struggle ensues. So too, many ecstatic or occult experiences are the result of defensive efforts by the ego to overcome various kinds of suffering, especially the loss of self-object support. Even when successful, these kinds of response deny an aspect of the ego's inner experience, namely "I have been hurt." The person and the person's culture may demand that we applaud these kinds of experience—for who are we to judge? But to applaud dissociative defenses and their contents, as clinicians, is highly unempathic. For as clinicians we are asked to judge continuously the veracity of patients' accounts, about the likelihood of this or that happening in their past and current lives, and about their version of human nature. As citizens we ought to tolerate the expression of religious beliefs of all kinds, unless they infringe upon basic rights of other citizens.

As clinicians does it help to agree with a patient's fantasies of adult functioning or those of the patient's family? Do we applaud these solutions when, arising from dissociated states and other attacks upon ego functioning, they deny the actual chain of events that eventuated in that solution? If, for example, an adult patient describes a childhood spent with a blatantly psychotic parent we are not surprised that our patient struggles with ordinary issues of reality testing and will necessarily reconstitute these struggles in the therapeutic relationship. Would it be helpful or therapeutic to ignore the patient's suffering as a child when the patient had to live in a world structured by incoherent beliefs?[44]

The answer to each question is surely no. As we saw earlier, where ambiguity persists and the need to know is great, we find the illusion of infinite depth. Lest we think this problem afflicts only humanists, in the next chapter we consider parallel events in the lives of scientists, beginning with the story of Percival Lowell, a distinguished American astronomer.

5. CANALS ON MARS:

EXPLORING IMAGINARY WORLDS

The primary task of the humanities, especially the arts, and theories *about* the arts, the work of humanistic research and scholarship, is to help us understand what it means to be a human being. This has two consequences for the issue of progress in the humanities. The first is epistemological. Anyone confronting this kind of gap, this murky area of not-knowing, whether labeled humanist or scientist or historian, will tend to project onto the noise one's favorite theory. The second is that in confronting this epistemological gap we often rely upon "great" persons whose intuitions about human being offer plausible, typically heroic, accounts of who we should become.

To be blunt, when we don't know how to demarcate known from the not-known we find multiple schools of thought, few or none of them compatible with the others. True, it might be that one of these great thinkers got it just right and that by following him we will understand exactly the issues at stake. And yet when we examine any one of these theories we do not find such a happy out-come. On the contrary, we find yet again disputes as to the precise meaning of what the word *language* means in Wittgenstein, for example, as we will see in the next chapter.

This is disappointing to those of us who want to feel that we inhabit a pro-gressive research tradition. As I noted in the beginning of this book, my own tradition of psychoanalysis is tipping toward oblivion. From an evolutionary point of view, that might be a good thing. It might be time to throw out a hundred plus years of observation and theory. I do not believe that is true at this time. Rather, my goal has been to acknowledge problems with Freud's explanatory theory, what he called his metapsychology. To do that and also to recognize the complexity of the task, I suggested that we can distinguish three discourses.

- The first kind of discourse includes rigorous studies of the biochemical processes of neural functioning, of neural transmission, and of the organization of neural structures in all species, from zebrafish to *Homo sapiens.*
- The second kind of discourse will include rigorous studies of the primary emotions (like hate, love, anger, and fear) that appear in all human beings.
- The third kind of discourse will include rigorous studies of specifically human feelings shaped by specific cultures and, perhaps, unique to specific persons.

It will be obvious to some that I have stolen this three-part set from American universities: part one corresponds to the sciences, part two corresponds to the social sciences, and part three corresponds to the humanities. While I acknowledge my debt to this traditional scheme, I am not ashamed. I take some comfort in the parallel between my brain-ego-self scheme, these three discourses, and the tripartite divisions that we find on most campuses. My comfort comes from the claim that the threesome has evolved over many hundreds of years into a reasonable division of labor on vastly complex problems.

Of course having distinguished three discourses we are left with the problem of trying to figure out just when one discourse should be replaced by another. Often it seems prudent to keep brain discourse separate from ego discourse and both of those separate from self discourse. Yet, interesting problems emerge exactly at the boundaries between one and another of these discourses.

Below we examine one such boundary dispute. This dispute occurs in the conduct of natural sciences when investigators push their ability to see objects of their inquiry that lay just beyond the scope of their instruments. In some of those moments in which the objects are not there, but are merely magnified or exaggerated, we see what Freud called "projection." Projection names that process in which our self—that is something within our minds and our wishes—ascribes to external objects features that originated within us. For Freud the most obvious example of such a projection was the image of God the Father of Judaism and Christianity.

In the previous chapter I said that we find numerous people, famous and not so famous, asserting that we can introspect, that is, look into our minds. In one very important sense this is true: only I, for example, know that I am hungry for a chocolate donut. Someone who knows me well and who has observed me on cold days might predict that I will heed the call of Dunkin' Donuts. But the undeniable fact is that I alone know precisely what I want today. It is easy to believe that I alone can know my feelings, thoughts, and fantasies. From this fact, it is not difficult to believe that we can discover essential truths about ourselves by looking inward, by self-analysis.

In contrast to faith in introspection, Freud's more important insight was we easily delude ourselves about how our mind works. Freud's genius was to mount a criticism of the universal fantasy that we can know our minds. The very idea of the unconscious mind is that a part of us, as Freud often said, a huge part of us, is not under our conscious control. Central to Freud's criticism of consciousness is that wishes, desires, and hopes and all the other elements that make up our libidinal selves can drive us to perceive things that are not really there. In this chapter I lean upon Freud's great insight to discuss how believing something turns into seeing something.

As always, the problem is how to find the breaking point between conviction and delusion, between hoping that something is true and believing that it is true. If my previous discussion about the error of searching for essences has been helpful, we should be wary of efforts to find the tipping point by looking inward. We will not find it through an introspective process. On the contrary, we will have to find ways or methods that can help us get down the road and toward something like plausible claims. Because we cannot look inward and rely upon what we find there, we must look outward, toward other people, and attend to their experiences of us. We need to find procedures that will let us make more reasonable claims and then find some ways to assess those claims.

This leads inevitably to talk about the "methods of science" or the "rules of inference." These are intimidating phrases. They may call to mind stern English teachers who rapped our knuckles about proper grammar. Or, they may call to mind stern professors who glared at our logical errors. I therefore replace these phrases with "a better understanding of our all-too-human wishes to experience ourselves in a certain way." I recount below how easy it is to fall into what some have called "pathological science." But that phrase is also too stern. It suggests that these are extraordinary faults, rare dysfunctions that afflict bad scientists. The evidence is that these are almost universal errors; in other words, they are ordinary human foibles. Thus, the following account is not a rogue's gallery of foolish people. Rather it is an account of intensely thoughtful people, most of them scientists, who went awry. We too can go awry even if we align ourselves with the sciences.

The first example of someone who gets it wrong is the American astronomer Percival Lowell. I recount his story and his fascination with what he called the canals of the planet Mars. Lowell's story exemplifies precisely what Freud called wish fulfillment overriding reality testing. I draw two morals from this account. The first is that Lowell believed incorrectly that seeing less was somehow seeing more. This is a fatal error for an astronomer to make. The second is that wishing something to be true, indeed finding it beautiful because it is true, easily turns into finding it true because it is beautiful.

The next set of examples of people who get it wrong comes from an important article by Irving Langmuir. Langmuir, an American scientist, investigated what he called "pathological science." From Langmuir and Richard Feynman, a distinguished American physicist, I draw lessons about the ethics of science.

Lastly, I turn to the English historian Thomas Macaulay, who wrote *History of England from the Accession of James II*, and to Herbert Butterfield, a historian who criticized the kind of historiography that Macaulay represents. Butterfield called this "the Whig interpretation of history," named after the Whigs, an English political party that claimed to represent the ineluctable rise of science and prosperity out of superstition and ignorance. Though given to disputation, Butterfield shows that in our drive to recover the narrative story, which we believe must course through history, we overly dramatize the past. We elect some as heroes whom we praise for having advanced historical destiny; we elect others as villains whom we despise for having retarded it. In doing so, we constrain historical accounts to narrow and even clichéd story lines about those who fought for and against a predestined historical outcome.

All told then, in this chapter I talk about some six or seven instances of thinking gone astray. At the end of the chapter I suggest two conclusions. The first is that if scientifically trained persons, some of them of eminence, can go awry, so too can children raised in a chaotic world. They cannot help but get it wrong. What they get wrong is the shape and rules of the self and social reality. This "getting it wrong experience" underlies a good deal of the suffering that brings people to seek out psychotherapy. This suffering derives from their efforts to make sense of the chaotic interpersonal world and their internal, chaotic feelings. Like scientists trying to magnify incoming data, children in these circumstances attempt to extract from that stream indications of danger and safety. In doing so, they make all the mistakes associated with pathological science.

The second conclusion is that when psychotherapists of all stripes search for the essences of their patients' suffering they also go astray if, having discovered that essence, they then elevate it to the level of a scientific metapsychology. Sometimes this leads them to search inwardly both the patient's experience and their own. When carried to an extreme and christened as a causal truth, this also becomes a form of pathological science because it forces the therapist to magnify limited bits of data and then extract from them meanings. In the psychoanalytic tradition, this search for meanings stemmed from Freud's original wish to model psychoanalysis upon the neurological sciences, to make it a form of breaking down complex symptoms into their parts. I have tried to show that this led him and others to mischaracterize psychoanalysis as a process of plumbing the depths. To put it more simply, Freud's commitments led him to see psychoanalysis as a process of magnification, aligned along a vertical axis. I have suggested that it's more accurate to see psychoanalysis as a process of exploration *between* persons and *between* persons and their environments, aligned along a horizontal axis.

For example, numerous clinicians have noted that the process of psychotherapy is akin to storytelling. There is much to recommend this version of psychotherapy: it links modern, secular therapy to the great traditions of religious cure of souls, to the ancient and universal feature of small groups and public myths, and to the development of self-identity through narration and stories. That religion has not disappeared from modern societies and that it has come roaring back into twentieth-first-century headlines suggests that religious narratives are not easily replaced by the narratives of the academy or the values of science.

It may be that religion will always be with us and that rationalism will never and perhaps should never replace it, at least in public discourse. Perhaps religion and religious narratives will replace those of the sciences. This, I hope, is impossible. We needn't look far to see the consequences for free inquiry and the advance of knowledge. If say the Southern Baptist Convention commandeered the National Science Foundation and the NIH and directed their resources to investigate the Holy Ghost, two bad things would occur. It would denature the religious core of Baptist thought, and it would destroy the National Science Foundation and the National Institutes of Health. That would appall most Southern Baptists, and it would remove any hope for scientific and medical advance. If religion and science are always with us, then there will always be at least two competing ways to view human beings.

VIRTUAL CIVILIZATIONS: PERCIVAL LOWELL AND THE MARTIAN CANALS

Virtual objects are those produced by microscopes, telescopes, and, Freud says, those seen by psychoanalysts. In the late nineteenth century new, more powerful telescopes seemed to reveal strange new worlds among the planets. Some astronomers observed dark lines on the surface of Mars that they called "canali" in Italian, or "canals" in English. Recalling our discussion of fancy, mimetic art and high art, we can say that canali evolved from mere virtual objects, lines seen in a reflection, to objects of fancy about which certain observers constructed complete narratives. That brilliant persons could do this suggests that the errors into which they fell were not ones of intelligence.

Giovanni Virginio Schiaparelli was a noted Italian astronomer esteemed for his studies of the planets. In 1877 Schiaparelli reported the presence of canali on the surface of Mars. Since he was working at the uppermost limits of the human eye and telescopes of his era, we do not fault Schiaparelli for naming the faint linear details he saw canali.[1] In *Planets and Perception*, William Sheehan describes Schiaparelli's gradual fascination with these lines and his increasing temptation to find in them more than was there. Schiaparelli subsequently named large areas on the planet "Lake of the Sun" and "Bay of Dawn" and called various canals "Indus," "Ganges," and "Hydraotes." We conclude that his wishful fancy had got the better of him. As Sheehan notes, by using such names Schiaparelli added to Mars and Martian studies "a considerable mythology."[2] Not only did these names evoke associations with great rivers of Earth, but Schiaparelli said that the canals were remarkably straight and geometric. Given the suggestiveness of these terrestrial names and the apparent geometry of the canals, they became inevitably the product of intelligent beings who had engineered them.

The stunning possibility of great civilizations upon Mars—for what else but an organized and intelligent race could construct and build such magnificent edifices?—fired the public's admiration. These ideas also fueled the imaginations of novelists and artists, who, in turn, increased the public's fascination with Mars.

As late as 1949, Chesley Bonestell (1888–1986), a renowned American illustrator painted a charming picture, "The Surface of Mars," complete with a perfectly straight canal.[3]

These claims also intrigued Percival Lowell (1855–1916), a Harvard mathematician who devoted himself to explaining these observations. A man of political power, education, and personal charm, Lowell brought to planetary studies "a ready-made interpretive framework for the observations before he can be said, properly speaking, to have put eye to telescope at all."[4] From these minuscule and fleeting visual phenomena, Lowell constructed an elaborate story of Martian civilizations, history, sociology, and so forth.[5]

Lowell was no fool. Benjamin Peirce called Lowell the most brilliant student he had ever seen at Harvard.[6] Indeed, Peirce invited Lowell to join the faculty. Born into a wealthy and famous family, Lowell was intimately tied to Harvard. At the age of sixteen, he first observed the heavens from atop the family mansion in Boston. From the next twenty years he searched for deeper religious ideals in the Far East, which he considered the "New World." Then, at the age of thirty-seven he learned that Schiaparelli was going blind and he received a book about the latest research on Mars. He became fascinated with the planet.

Just as Lowell looked to these new discoveries as sources of new, practical knowledge, other late Victorians became fascinated with the great pyramids as sources of untapped, transcendental wisdom. For example, C. Piazzi Smyth, the astronomer royal of Scotland, wrote a major work in 1880 on the Giza pyramids that he entitled *Our Inheritance in the Great Pyramid*. Charles Taze Russell, who founded the Jehovah's Witnesses, used elements from Smyth's book in his study *Thy Kingdom Come* (1890). In that book he argued that these uncanny measures were evidence for his claims that the ratios and measurements within the great pyramids were signs of God's plans for the human race, that these marvelous geometric achievements prefigured the coming of the Christian deity.[7]

In a similar way, Lowell conceived of the "philosophy of the cosmos," a scheme based on the idea of evolution. Charles Darwin, Alfred Wallace, Thomas Huxley, and Herbert Spencer had, it seemed, advanced other sciences using evolutionary models. So too Lowell would describe cosmic evolution, citing new evidence from astronomy. Sheehan cogently cites a passage in which Lowell described his beliefs about the planets: "Each body under the same laws, conditioned only by the size and position, inevitably evolves upon itself organic forms."[8] This lovely sentence is a precise statement of Lowell's faith and wish to discover that an overall "law" governed the appearance of life on earth and on all the other planets. We note that this wish is contrary to Darwinian theory, for Lowell does not underscore the central role of chance and randomness as they shape evolutionary paths; rather, it replaces the Genesis narrative of the dome of heaven with a modern narrative about a fixed and universal pattern of all the heavens.

In a stinging criticism of Lowell's 1908 book about Mars, Eliot Blackwelder, a distinguished American geologist, noted that Lowell used bits and pieces of geological history to maintain the story that the Earth and Mars shared a common evolutionary history. For example, Lowell asserted that both Earth and Mars show the rise and then decrease in the size of the oceans. If this were so, Earth's future would be seen in Mars's current desolation. Yet, Blackwelder asserts, "The truth is that there have been fluctuations of land and sea throughout recorded geologic history, and these changes show no general tendency."[9]

Shaped by his sentiments and using a telescope borrowed from Harvard, Lowell set out for Arizona to create an institution dedicated to Mars. In 1894, using his own money, he established an observatory at Flagstaff, Arizona, and in 1895 he wrote his first book on Mars.

Lowell's sketches of the canals of Mars varied from faint pencil lines drawn during an observational period to much more elaborate compendium drawings composed over a lengthy period of drafting.

Of special interest to Lowell were paired canals, that is, lines that seemed to run parallel to one another for many hundreds of miles. While Lowell acknowledged that we could not see the actual canals using telescopes of his era, we could see the broad sweeps of irrigated land that lay on either side of them.

The authors of the article on Mars in the eleventh edition of the *Encyclopedia Britannica* (1910) question Lowell's claims about the canals. They refer to "psychological optics" rather than astronomy: "The difficulty of pronouncing upon [the canals'] reality arises from the fact that we have to do mainly with objects not plainly visible (or, as Lowell contends, not plainly visible elsewhere). The question therefore becomes one of psychological optics rather than of astronomy."[10]

Using Freud's description of wishful thinking, Sheehan shows how Lowell's profound romanticism and his passionate "will to believe" animated his drive to confirm the reality of life on Mars. The same passion caused Lowell to overlook many refutations and objections to these claims. These refutations are a mixed lot, composed of logical problems, theoretical objections, thought experiments, and, most telling, observational data. Taken as a whole they weigh against Lowell's claims. Chief among these refutations were the following:

1. As early as 1877, Schiaparelli noted that the canals varied dramatically in type, location, and shape, sometimes disappearing altogether in the course of the Martian year.

2. As he continued his drawings Schiaparelli's renditions of the canals became straight lines. Yet, if they were actually the shortest distance on a globe—as the Martians would surely build them—they ought to appear as arcs of a circle.[11] (Lowell corrected this in his later books.)

3. Other astronomers who also claimed to see canals failed to locate them in the same region of Mars as did Lowell.

4. Schiaparelli was colorblind. Astronomers who were not colorblind noted that colored areas and the boundaries between them were soft, not the hard and rigid lines Schiaparelli recorded and that the canal theory demanded.

5. When observations accumulated slowly about gigantic mountains on Mars, Lowell consistently denied their validity. For his theory—that intelligent beings had constructed canals for irrigation—required them to do so on a flat planet. Therefore, when confronted by new observations, distinct but pertinent to the canal controversy, Lowell felt impelled to reject them.

6. The irrigation scheme that Lowell claimed to observe upon the Martian surface required water from the planet's northern pole irrigate the southern regions and water from the planet's southern pole irrigate the north-

ern regions. This manifest absurdity did not go unnoticed by critics, but was happily overlooked by a credulous public thrilled by Lowell's sensationalism.[12] (Rising to the occasion, both Lowell and Schiaparelli argued Martian sociology, the former favoring a view of Martians as oligarchic and elite, the latter suggesting that Martians had formed a "socialist confederation.")[13]

7. Neutral observers, using larger and better telescopes, available after 1877, failed to see the canals. Indeed, examining the surface of the planet they discovered that apparently smooth or linear details exploded into increasingly complex and variegated entities. The lines observed in smaller telescopes dissolved into dots, patches of color, and edges of one color shading into another.

This last item is the most compelling argument against Lowell's canal theory. When a host of astronomers advanced it, Lowell dismissed them out of hand. A standard value or rule of astronomical observation is that seeing more is seeing better. When we apply this rule to these new observations of Mars, this rule requires us to doubt Lowell's claim that canals existed on its surface. Lowell retorted that the canals appeared only to those looking at the surface with less powerful magnification. If this logic were allowed to stand, then it effectively meant that the very best viewing would be with no lens at all.[14]

In an article for the *Atlantic Monthly* in 1895, Lowell explained his discovery:

> The result is that the whole of the great reddish-ochre portions of the planet is cut up into a series of spherical triangles of all possible sizes and shapes. What their number may be lies quite beyond the possibility of count at present; for the better our own air, the more of them are visible. About four times as many as are down on Schiaparelli's chart of the same regions have been seen.[15]

Perhaps the most telling criticism of Lowell's personality is that he collected adherents, not fellow scientists. When more and more evidence amassed contrary to his profound wish to discover life on Mars, Lowell refused to acknowledge it. In this sense, as Sheehan says, "Lowell cannot be called great."[16] On the contrary, using Freud's definition of illusion, a belief grounded on wishfulness and impervious to refutation, Sheehan demonstrates how Lowell refused to give up his lifelong hopes about the Martian canals. Sheehan follows Freud directly in his history and notes how "the will to believe" can overrule both intellect and perception.

As I noted in my study of the occult, Sir Arthur Conan Doyle is a precise replica of Lowell and of Freud.[17] For while Freud enunciated the principles of rationality with which Sheehan agrees, in his private life and typically following the loss of persons important to him, Freud regressed to wild speculations about occult psychic powers, including numerology and telepathy.

Lowell and his followers used science and its instruments to validate their mythic wishes and then literally blinded themselves to further discovery. When he "stopped down" the aperture on the better telescopes, thereby reducing their magnification, Lowell demonstrated a typical response anytime a mythic system is challenged. Rather than open up discourse, we attack those who raise the questions or exclude them from the debate. This is not to condemn Lowell or to suggest that he was not a good scientist in other moments. The evidence is that he was an honorable person in many ways and upon other topics made genuine scientific advances. However, answering as it did so many of his boyhood conundrums and fulfilling his dearest wishes, the canal myth was too beautiful to abandon. When fellow scientists contested his observational claims, Lowell retreated.

Some might suggest that Lowell is a special case of a brilliant, but disturbed mind; that he had a hidden psychosis. While not impossible, this explanation for Lowell's persistence begs the question since there is no independent evidence that Lowell was delusional. On the contrary, Lowell was an admirable person with no overt thought disorder. A more parsimonious claim is that Lowell saw real things when he looked through his telescopes, namely the fuzzy and inconclusive images projected there by the instrument's lenses. Having found the precise level of magnification required to create these fuzzy images, Lowell remained fixed there. By remaining fixed within that station, and by literally connecting the dots, he drew lines that he named canali. Using these completed maps Lowell believed that he could show evidence of intelligent design. Once we agree that only intelligent beings could have designed and built such immense structures, Lowell asserted that we can deduce likely truths about the civilizations that carried out these feats.

In his 1906 book on Mars, Lowell employs his logical gifts to demonstrate that his observations of canallike structures must be canals: "For we find ourselves confronted in the canals and oases by precisely the appearances *a priori* reasoning proves should show were the planet inhabited. Our abstract prognostications have taken concrete form. Here in these rectilinear lines and roundish spots we have spread out our centres of effort and our lines of communication. For the oases are clearly ganglia to which the canals play the part of nerves."[18]

By pursuing this mythical form of reasoning, Lowell removed himself from among those who examine the world of affordances, which is the world of high art. For as Lowell says, he discovered canals and oases (yet another water reference) precisely where his imagination had wished them to be. Reasoning that merits the title "a priori" is logical deduction conducted within a system of assumptions, axioms, and rules of combination, for example, like those of

Euclidian geometry. In those systems one can indeed deduce new, valid propositions because a fixed set of rules governs the production of well-formed formulas. Each well-formed formula counts as a valid proposition whose truth is guaranteed by its grammatical correctness.

In his essays on method, Descartes, a great mathematician, noted that this form of geometric reasoning occurs without reference to observations of the natural world because the objects of inquiry are themselves abstract and defined by the rules of the system. Objects of mathematical reflection and induction belong to what Karl Popper called World III.[19] World III corresponds to that real, invisible set of ideas, concepts, and "structures," especially language, to which some forms of self-reflection give access. Unlike the second world, which seems wholly under the sway of idiosyncratic experiences, the third world reveals permanence and coherence that distinguish it from fantasy. Western examples include the notion of transcendental mind, like that evident in Parmenides' poem about "Being," the Greeks' discovery of irrational numbers, the notion of Justice, and all the logical sciences. We note that genius typically refers to mastery of these kinds of objects. Music, logic, mathematics, and chess are all activities that deal with World III entities.

By masking his empirical observations of Mars as a set of logical truths, as if they were World III entities, Lowell returned to the ranks of those who write fancy. Fancy is fundamentally different from the world of art. Fancy is driven by wishes; it easily discards the constraints of nature and of logic. For example, not-very-good science fiction describes time machines that a backyard tinker assembles in his or her basement workshop. From there, he or she can have numerous adventures that resemble daydreams in their wishful themes and illogicality. These can be fascinating. But their fascination derives from their resonance with our fantasies, or what Ruskin called fancy. Because they are not based on observation they cannot tell us anything new about the world of real persons acting in real time.

Once Lowell had entered the rank of myth and fancy, he was able to compose detailed narratives about life on Mars without the constraints of observation, or history, or any other kind of research. Lowell's fascination with the (illusory) canals of Mars may demonstrate his need to find in science semblances of a coherent-self objectified, as it were, by the projection of organized and reasonable life onto the distant planet. Like William James who defended religion in *The Varieties of Religious Experience* (1902),[20] Lowell defends the wish to believe that we are not alone. One might see both Lowell and James in retreat from the awfulness of "the new sciences," especially Darwinism and its philosophical destruction of the argument from design for the existence of God.

Looking through a certain sized lens, Lowell perceived virtual images, namely, faint linearlike shapes on a red, disklike object called Mars. He could not perceive visual affordances because he, a medium-sized mammal, was *not* located within an actual world in which he could perceive affordances of any kind. As Gibson says about dreams, within a dreamscape I may believe that I can investigate the objects around me, but since they are not real, they offer no affordances. I cannot walk around them, say, to gain yet another perspective on them. I can *imagine* walking around them; however, this gives me no new information, except to learn, perhaps, how my fancy works. I can imagine walking around Paris, and I might enjoy that sojourn, but I could not learn anything new about the real Paris.

This brings us back to the distinction between seeing a real landscape and imagining that one sees a landscape, for example, as in dreams. As I noted above, Descartes relied upon his account of visual illusions to support his claim that the mind and body are wholly separate entities. For example, Descartes talks about seeing in dreams: "The visions which come to us in sleep are like paintings."[21] This seems plausible: we cannot walk behind (most) paintings just as we cannot walk around or behind dream scenes. However, even Descartes's clever insight offers too strong a model, because we cannot approach a dream scene the way that we can approach a real painting in a real gallery. And as we noted in the previous chapter, it seems very likely that we understand that something is real because of the millions of unconscious calculations we make as we move through the world. For example, we notice the change in the visual angle as we approach a real painting in the real world.

The work of high art is not magic. Any clinician working with borderline personality disordered (BPD) patients knows well the experience of being just a bit irritated with a patient and then being nailed by the patient's exact, laser beam of suspicious examination. While BPD patients often overread and over-interpret other persons' actions, thus producing many miserable moments for those around them, they do not hallucinate. Like the woman of genius whom Henry James described, they are highly attuned to nuances of possible meanings within a social exchange.

Borderline patients extend their reading of other people's motives into a stereotyped story: attack, defense, and attack again. Because everything appears dangerous to them, they magnify signals of potential danger and in so doing distort their vision. Like putting a magnifying lens over someone's face, ordinary blemishes and faults appear cavernous and disgusting. Borderline patients hone in on any and all faults; needing to find the danger hidden just beneath the therapist's face, they confront the therapist again and again. The BPD patient drills into parts of the other, searching for the secrets they know must be there and dread finding.

BPD persons are attuned to others but they often go wrong. Once they settle upon a given percept, they respond to it as Lowell responded to the canali. Like anyone caught up in a moment of passionate sureness, they pursue a kind of pathological form of knowing. We are all liable to fall in to this error, even persons trained in the rigorous sciences.

PATHOLOGICAL SCIENCE: THE LIMITS OF VISION

Lowell's story fits the pattern of science that Irving Langmuir called "pathological."[22] Langmuir, a Nobel Prize–winning chemist, discusses five instances of science gone awry. Among these cases are his investigations of an alpha-particle experiment and a delightful account of his visit to Professor J. B. Rhine, who claimed scientific evidence for extrasensory perception (ESP). Alongside these two firsthand accounts, Langmuir cites similar stories about "N-rays," "Mitogenic rays," and the "Allison Effect." Some scholars take offense at the word *pathological* since it seems to verge on moralizing about colleagues who were not fraudulent.[23] With the exception of J. B. Rhine, this seems to me a fair assessment. However, Langmuir used the word in a clinical sense and I will follow him in that usage.

In many cases of pathological science, we find errors of judgment based on attempts to interpret very faint—or nonexistent—visual data. For example, Langmuir describes his investigation of "alpha particles," entities that Bergen Davis and Arthur Barnes, two American scientists, claimed they could discern by very carefully counting "scintillations" on a screen observed under almost complete darkness, using a microscope (in other words, they tried to evaluate virtual images). Langmuir visited their laboratory and tried to emulate their method. He found that he could not see as many scintillations as Barnes could. Asserting that the frequency of scintillations depended upon the current applied, Barnes and Davis claimed that every time they altered the voltage they counted more or fewer scintillations. If this were true, it would bring important experimental data in support of Niels Bohr's theory of the atom.[24]

After hearing an enthusiastic talk from Davis on their startling discoveries in 1930, Langmuir and a colleague, C. W. Hewlett, an expert on Geiger counters, visited Davis and Barnes in Davis's laboratories at Columbia University in New York. After getting their eyes adapted to the darkness for half an hour, the two visitors counted scintillations on the apparatus designed by Barnes and Davis. Langmuir and Hewlett counted around 60 (or one per second), while Barnes counted 230 on his first attempt.

After many observations of their technique—in which an assistant altered the voltage and Barnes counted the scintillations produced by these discrete voltages—Langmuir noticed that Barnes could see the assistant sit back whenever there was *no* voltage. This suggested to the person counting scintillations that that there should be fewer. And, indeed, when the assistant sat back, Barnes found fewer. To test his speculation, Langmuir asked the assistant to alter the voltage, but remained fixed at the machine, letting Barnes (in this case) continue to count scintillations: "Whether the voltage was at one value or another didn't make the slightest difference."[25] Barnes's counting was right half the time, and wrong half the time; the count decreased with the voltage off about half of the time. Confronting Barnes and Davis, Langmuir showed them that being right half the time, when there are only two end-states possible, meant that they were not counting anything. It was all mere guessing. Both scientists were shocked by this conclusion; Davis immediately made up an excuse: "He had a reason for not paying attention to any wrong results," Langmuir noted.

It took eighteen months for Barnes and Davis to withdraw their claims about alpha particles, noting in their published retraction that counting scintillations on a screen involves threshold phenomena, easily influenced by the observer. Langmuir's report suggests a more stringent assessment that Barnes and Davis were so influenced by their wish to find evidence that they violated ordinary experimental rules. For example, Barnes claimed that he always counted scintillations for two minutes, that is, 120 seconds. Langmuir timed Barnes and found that these viewings varied between 70 seconds and 115 seconds, or between 58% and 96% of the alleged period: "But he counted them all as 2 minutes, and yet the results were of high accuracy!"[26] If the alpha particles were real entities produced by some as-yet-unknown physical event, such variable sampling should produce a variable count, not the unvarying results that Barnes obtained. This could happen only because Barnes wished to find a fixed number over a fixed period of time. Both the number of scintillations and the period of time were subordinated to his fantasy.

Similar forms of error occurred in attempts to discern the presence of "mitogenic rays." According to Alexander Gurwitsch (1874–1954), a Russian biologist, mitogenic rays were given off by all living things, from plants to people. To discern evidence for these rays Gurwitsch and his followers examined very faint images on photographic plates. However, as in the other stories, these plates had very slight—or no—differences from nonexposed plates. Langmuir notes, "If you looked over those photographic plates that showed this ultraviolet light you found that the amount of light was not much bigger than the natural particles of the photographic plates."[27] This meant, again, that investigators had to examine each plate with attention to ill-defined details that might be the effects of random processes, such as background radiation or mere noise. The result was that about half of the people who examined these plates saw evidence of the alleged rays and half did not. As I will note in the discussion of *High Anxiety* in chapter 7, this is precisely the state in which Dr. Thorndyke's assistant found himself as he magnified the tiny bit of visual data some thirty thousand times.

Given that half of the viewers will see evidence of the mitogenic rays, it's not surprising that those who see what is so subtle will be convinced and will find reasons to continue the work begun by the master. Thus, some Russian scientists have continued the search for ultralow emission evidence of mitogenic rays. At a conference in 2004, numerous papers were presented on topics like "a holistic organization of cells and multicellular systems as revealed by studies of the ultra-weak photon emission":

> In spite of a domination of reductionist approaches in modern biology, the data are accumulated supporting a view of the living cells and multicellular collectives as unified continuums (fields). A historical milestone in establishing this approach was a discovery of so-called degradational ultraweak photon emission (UPE) by Gurwitsch' school [sic].[28]

Langmuir cites the case of "N-rays" and their alleged discovery by a French physicist, René Blondlot, around 1903.[29] On analogy with X-rays, Blondlot claimed that N-rays could penetrate some ordinary substances to great depths, such as many inches of aluminum, but could not penetrate thin sheets of iron. Given the spectacular discovery of X-rays by Wilhelm Conrad Roentgen in 1895, the world was ready for another astounding revelation of unseen powers.[30]

An American physicist, Robert W. Wood, investigated Blondlot's claims and observed his laboratory. Wood noted that Blondlot claimed to have a supersensitive skill in discerning the effects of N-rays upon various substances: "At that time Blondlot was using a spectroscope fitted with an aluminum prism to measure the refractive indices of N rays" to within three significant figures on a barely illuminated scale.[31] Watching Blondlot carry out his work with the prism, Wood decided to test his colleague. Under cover of darkness, he removed the aluminum prism, the supposedly critical part of the instrumentation. Wood asked Blondlot to repeat the experiment; he did so and obtained the same results. Wood published his story and this killed research on N-rays, outside of France.

Blondlot and others investigated N-rays and discovered, so it seemed, that they could discern the presence of N-rays on very, very faintly illuminated paper screens: "To make sure that you are seeing the screen of paper you hold your hand up and move it back and forth. And if you can see your hand move back and forth then you know it is illuminated."[32] In other words, to test whether or not there was sufficient illumination, these investigators used their common sense: if they could see their hands move, this proved that they were able to discern visual data. Repeating these methods, a German scientist discovered that he believed that he could see his hand regardless of whether or not it was in front of or behind the paper screen. This is puzzling, since it would seem that the paper should hide one's hand if it were behind it. Langmuir notes, wryly, that you see your hand there because you know it is there.

ESP AT DUKE: THE STORY OF J. B. RHINE

Following his experience with Barnes and Davis and their elusive alpha particles at Columbia, Langmuir traveled south to visit J. B. Rhine at Duke University in 1934. He gave Rhine a full account of the self-delusion he had discovered in Barnes and Davis's efforts to count scintillations. Rhine did not see how his efforts to study extrasensory perception (ESP) paralleled the errors committed by his physics colleagues. Rhine's method, immortalized in the opening scenes of the movie *Ghostbusters* (1984), involved a "transmitter" and a "receiver." Harold Ramis and Dan Ackroyd, the screenwriters of *Ghostbusters*, make the ESP experiment entirely fraudulent since Professor Peter Venkman (Bill Murray), an alleged expert, carries out investigations of ESP, using a painful shock designed to punish "bad" ESP guesses. To seduce a lovely coed, Dr. Venkman—who claims a PhD in psychology and a second PhD in parapsychology—rewards her errors and punishes the correct guesses of an obnoxious young man.

In Rhine's usual setup (which had no shock device) the transmitter looked at a card, thought about it, and the receiver tried to surmise the card. Rhine used a deck of five suites—Circle, Cross, Wavy Lines, Square and Star—of five cards each. Rhine claimed that after millions of trials, the average number of cards identified correctly was seven, far higher than five, the number we'd expect to find if the process were genuinely random and the receiver were guessing.

Chatting about these extraordinary results, Langmuir says that Rhine told him "people don't like me," meaning, apparently, that some people gave artificially low ESP readings in order to spite him. Rhine locked up these low scores in a file cabinet. Langmuir challenged Rhine to publish all of the ESP data and to include the low scores (from the spiteful, anti-Rhineans) alongside the high scores. If those were added to the high scores the average of correct guesses would likely decrease to around five per twenty-five attempts, precisely what chance would dictate. Rhine did not publish the low scores and with that subterfuge forfeited any hope of plausibility. Even if Rhine obeyed the rules of ordinary scientific practice, his reasoning was unintelligible since he offered neither a theory as to how ESP worked, nor why it seemed to work equally well when subjects predicted the run of cards on the *following* day. As Langmuir notes, one can imagine that with intense thinking a person might transmit some kind of supersubtle brain waives picked up by a supersensitive receiving subject. Yet, Rhine reported that his best subjects could predict a card sequence twenty-four hours *later*. What possible mechanism could account for that feat? Looking for the proper occult term, Langmuir mumbles: "You record them [the guesses] and then you look them up and see if they check, and that's telepathy, or clairvoyance rather. Telepathy is when you can read another person's mind."[33] While we might quibble about the proper terms, it is difficult to see how these two supposed events are similar to one another. It is even more difficult to imagine a brain or mental mechanism operating in both circumstances.

Crucial to understanding Langmuir's insight about pathological science is his comment about ESP claims: "All of these things are nice examples in which the magnitude of the effect is entirely independent of magnitude of the cause." This marks ESP claims as radically unlike typical claims about causal relationships in the real world. For in most cases we find that increasing the size of the cause, that is, increasing its power, should increase the size of its effects. But in Rhine's experiments we do not find any such expectable correlation. As Langmuir says throughout his papers, this failure of expected relationships becomes one of the defining characteristics of pathological science. Contrary to normal accounts of causal linkages, we find no discernable correlations between size of cause and size of effect in these accounts. We do find a fixed correlation between the scientists' wishes to find causal connections and the size of their claims. There, we find a typical pattern.

As Langmuir notes, those who believe in the claims, those who have wedded themselves to the feelings and enthusiasm of the discoverer, can easily make up clever (or semiclever) excuses for experimental failures. For example, contemporary enthusiasts repeat Rhine's defense of less-than-average runs of ESP cards. They write that apparently some people "dislike ESP. Even though the subjects were consciously trying to achieve good scores, they scored lower than chance. An unconscious factor seemed to come into play here. Experimenters have found they can predict higher scores for some groups (for example, those who are interested and relaxed), and lower scores for other groups (those who show fear, negativity, or boredom). The factor of missing-ESP indicates why ESP data is unreliable."[34]

This defense of Rhine's methods raises question-begging to an art form. This paragraph can make sense only if one knows that ESP is a genuine skill that all people possess. If it's true that we all have this skill, then any score that is lower than the five per twenty-five average must be the result of occult processes inhibiting the free expression of ESP. These occult processes are "unconscious," which means that neither the subject nor the experimenter can assess them directly. Hence, when subjects were excited about ESP and wished to demonstrate their gift, and failed, there must be countervailing wishes or forces that impeded them. If the experimenter cannot identify "fear" or "boredom" then he or she can deduce the presence of "negativity."

A more plausible explanation for these failures is the ordinary law of averages and regression to the mean. If we excise the fundamentalist assertion that ESP must be real, then like the N-rays, it too disappears. Langmuir's criticism did not stop Rhine from publishing a book several years later, *New Frontiers of the Mind: The Story of the Duke Experiments*. There Rhine attempted to distinguish himself from parapsychology and claimed that he followed strictly experimental methods.[35]

Langmuir summarizes characteristics of "pathological science":[36]

1. The maximum effect that is observed is produced by a causative agent of barely detectable intensity, and the magnitude of the effect is substantially independent of the intensity of the cause.
2. The effect is of a magnitude that remains close to the limit of detectability or, many measurements are necessary because of the very low statistical significance of the results.
3. There are claims of great accuracy.
4. Fantastic theories contrary to experience are suggested.
5. Criticisms are met by *ad hoc* excuses thought up on the spur of the moment.
6. The ratio of supporters to critics rises up to somewhere near 50% and then falls gradually to oblivion.

Returning to the story of Percival Lowell, we find a set of errors parallel to those Langmuir describes in his diagnostic essay. Most striking is the precise ways that Lowell's errors replicate the first two points. It was only by selective, highly individual attention to the faint images of canali that Lowell and others could see them. Any variation in focus or in the selection of magnification destroyed the "virtual" image and, therefore, the entire story of Martian civilization. The third point, replicated when Lowell claimed great accuracy in his observations, contradicts the second, that he was working at the edge of discernability. It cannot be that a given signal (an image, in this case) is both extremely faint and extremely clear, that is, of such magnitude that one can claim great accuracy in measuring it.[37]

To Langmuir's list we can add a seventh item: intellectual honesty.

"CARGO CULTS" AND THE ETHICS OF SCIENCE

While it's true that most courses called "Ethics" are taught in humanities departments, especially philosophy and religion, it does not follow that humanists understand ethics in ways that are superior to scientists, for example. On the contrary, in most ethics courses the focus turns on exquisite analyses of the meaning and use of moral terms, such as "good" or "values" or "justice."

Richard Feynman, famous both for his scientific work in physics and his popular explication of scientific reasoning, made a similar distinction between what he famously called "cargo cult" science and authentic science. The first derives from well-known studies of South Sea Island peoples who, after World War II, believed that if they replicated the behaviors of the American crews who landed all those cargo-laden airplanes during the war that the airplanes would return: "They're doing everything right. The form is perfect. It looks exactly the way it looked before [during the War]. But it doesn't work. No airplanes land. So I call these things cargo cult science, because they follow all the apparent precepts and forms of scientific investigation, but they're missing something essential, because the planes don't land."[38] What are they missing? Feynman's answer is that scientists should strive for integrity, "a kind of leaning over backwards. For example, if you're doing an experiment, you should report everything that you think might make it invalid—not only what you think is right about it: other causes that could possibly explain your results; and things you thought of that you've eliminated by some other experiment, and how they worked—to make sure the other fellow can tell they have been eliminated."[39]

This is an ethical demand. It is the opposite of arguing *for* one's favorite idea. It requires one to examine one's claims from all possible angles, to look at the phenomenon from as many different viewpoints as possible. This is the rule that Lowell, and others who fell into wish fulfillment, failed to observe. When he rejected the findings of others who used different instruments, Lowell championed only what he thought was correct about his observations of the canali. In contrast to Lowell, Feynman's answer is a kind of ethic directly opposed to polemics, to political argument, to legalisms, and to rhetoric. The goal of authentic scientific practice is "to give *all* of the information to help others to judge the value of your contribution; not just the information that leads to judgment in one particular direction or another." To demand that we give all the information—the good, the bad, and the ugly—plenty of renowned scientists fail. They fail at this because they believe that their self-esteem, their funding, and their fame rest upon making heroic discoveries.

In the earlier portions of this chapter we focused upon the activities of scientists. This may seem unfair since we have not talked about the errors of humanists. I now correct that mistake. For while I have discussed those who use the words *pathological science*, I have not shared their tone of condemnation. On the contrary, just as scientists can fail, we humanists fail. One way we fail is evident in authorities who write national histories and in the work of psycho-therapists who write personal histories. In both kinds of narrative we are tempt-ed to write a drama, to tell the story of the coming to be of either a nation or a human being. We are tempted to write some version of heritage and not engage in the rigor and tedium of history. These are vastly different tasks. By its very nature the writing of heritage is an emotional event in which narrator and read-er are caught up in something akin to a religious experience.

If, for example, we are British (or severely Anglophilic) the prose of Thomas Macaulay and Sir Winston Churchill provokes deep feelings about our national character, our British heritage.[40] While they are separated by about one hundred years, the tone of their histories is identical: each is a dramatic recount-ing of the divine forces that led to the creation of the English Empire and, in Churchill's account, to the rise of the United States, England's greatest offspring.[41]

To be compelling, a narrative must evoke from us a sense of danger, struggle, and yet possible escape. We must grasp immediately what is right and what is wrong. The quickest way to do this is to show us a crime in process and then indicate that an even larger crime is about to be committed unless the forces of good (always outnumbered and seemingly overwhelmed) intervene. A modern American master of this is George Lucas, author and creator of the *Star Wars* sagas. The same formula appears in his lesser-known film, *Willow* (1986).[42]

In this film—a retelling of the hero myth with elements of the Christ sto-ry—the opening act is the birth of an innocent baby in a dark, volcanic land-scape. The mother screams and a baby girl emerges, bearing a special mark. The evil queen, Bavmorda (Bav = Bad? Morda = Mordant, Murder?), seeks out newborns who bear the special mark because, like Oedipus's father, she has learned through prophecy that a certain child will rise up to defeat her. Seeking to destroy the child she vows, "This child will have no power over me." Luckily, as we see the evil queen approaching the newborn, a kindly old midwife saves the baby. Too late, Bavmorda sees that she's been tricked and orders the mother to be killed. In the next scene violent horsemen rampage through a peasant village with "death dogs" seeking out the special baby. We see them abuse innocent people left and right, then murder one in front of our eyes. The midwife huddles with the baby, attempting to secure it against these monstrous powers. The next day, still pursued by the guards and the death dogs, she remembers (it seems) how Moses was saved and she creates a kind of raft, places the baby on it, and suffers it to be swept down the stream. Hearing the dogs she lures them to follow her, and as they attack her the baby drifts off to safety. The shooting script tells with whom we should identify: "From the baby's view, as it floats down the river, WE SEE the dogs kill the midwife beyond the distant trees."

We note the parallel to the story of Moses (Exodus 2:2–10) and the birth of Jesus (Matthew 1:18, 22–25 and Luke 1:26–38). In each story a monstrous, earthly power seeks to destroy the hero-rescuer-Messiah whose conception, birth, and destiny are divine. Within the framework of myth, contemporary authors typically find ways to link all of these events into a single narrative.

THOMAS MACAULAY AND ENGLISH DESTINY: HISTORY AS GRAND NARRATIVE

Born in October 1800 and the son of Zachary Macaulay, a former African colonial governor, Thomas Macaulay emerged as a brilliant opponent to slavery even while a young man. At age thirty he entered politics as a member of the Whig Party, roughly the liberals of their time. A brilliant and accomplished classicist and conversationalist, Macaulay identified himself with his father's zealous attack upon the institutions of slavery and with the progressive spirit of the Whigs, whom he claimed represented the true spirit of British history. Like Abraham Lincoln, an American Whig, Macaulay believed that progressive policies, land development, and an attack on landed privilege and particularly upon slavery were essential to making manifest the destiny of each nation.[43]

Macaulay's Introduction to *The History of England from the Accession of James II* begins with an overture that sums up the dramatic story he wishes to tell about the coming to be of England.[44] Without naming God, Macaulay makes it clear that he will trace the origins and destiny of the English. For England, in his view, was born midst struggle between forces that wished to prohibit the growth of liberty:

> I shall trace the course of that revolution which terminated the long struggle between our sovereigns and their parliaments, and bound up together the rights of the people and the title of the reigning dynasty. I shall relate how the new settlement was, during many troubled years, successfully defended against foreign and domestic enemies; how, under that settlement, the authority of law and the security of property were found to be compatible with a liberty of discussion and of individual action never before known; how, from the auspicious union of order and freedom, sprang a prosperity of which the annals of human affairs had furnished no example.[45]

This combination of freedom and an intelligent Whig government made England richer than any other empire and "not less splendid and more durable than that of Alexander." As he makes clear at the end of the introduction, Macaulay sees his task as telling a single great story, a great narrative of the rise of England: "The events which I propose to relate form only a single act of a great and eventful drama extending through ages, and must be very imperfectly understood unless the plot of the preceding acts be well known." This makes his account fascinating and enjoyable—a bold and persuasive recounting, like a movie version of a complex event. Thus concerning the "Conversion of the Saxons to Christianity" Macaulay says, "The darkness begins to break; and the country which had been lost to view as Britain reappears as England. The conversion of the Saxon colonists to Christianity was the first of a long series of salutary revolutions." Though he will later castigate the church for its excesses and for choosing at times to deny what he sees as progress, Macaulay can celebrate the monastic tradition that kept learning alive. Again: a dramatic scene of the few forces struggling to maintain decency and learning against the dark, regressive forces: "The Church has many times been compared by divines to the ark of which we read in the Book of Genesis: but never was the resemblance more perfect than during that evil time when she alone rode, amidst darkness and tempest, on the deluge beneath which all the great works of ancient power and wisdom lay entombed, bearing within her that feeble germ from which a Second and more glorious civilisation was to spring."

In the same cinematic sweep, Macaulay describes invaders from Denmark and Scandinavia who were "distinguished by strength, by valour, by merciless ferocity, and by hatred of the Christian name. No country suffered so much from these invaders as England. Her coast lay near to the ports whence they sailed; nor was any shire so far distant from the sea as to be secure from attack." Civilization had begun to rise but was stuck down by the Danes.

Concerning the Norman conquerors, Macaulay excludes them from the title of English: "The Conqueror and his descendants to the fourth generation were not Englishmen: most of them were born in France: they spent the greater part of their lives in France: their ordinary speech was French: almost every high office in their gift was filled by a Frenchman." His prose is vigorous and the story is one of bare escape, for the essence of England and its later greatness depended upon the lucky (or foreordained) accident that the French finally produced a bad king: "Had the Plantagenets, as at one time seemed likely, succeeded in uniting all France under their government, it is probable that England would never have had an independent existence."

Macaulay uses another device to hammer his point home. He refers to the origins of great rivers in small, hidden, and rude parts of distant mountains:

"The sources of the noblest rivers which spread fertility over continents, and bear richly laden fleets to the sea, are to be sought in wild and barren mountain tracts, incorrectly laid down in maps, and rarely explored by travelers. To such a tract the history of our country during the thirteenth century may not inaptly be compared. Sterile and obscure as is that portion of our annals, it is there that we must seek for the origin of our freedom, our prosperity, and our glory."

This biblical sweep amplifies Macaulay's feelings that his story is the account of the inevitable rise of a Just Nation, a True English Nation, whose destiny was to increasing glory. Macaulay's language is similar in tone to the Hebrew Bible, especially the prophets like Isaiah, who tells us that when the Messiah arrives "every valley shall be exalted, and every mountain and hill made low, the crooked straight and the rough places plain" (Isaiah 40:4). Like that text and the Handel oratorio into which it was placed, Macaulay makes a prophetic claim. Like the ancient English, like the baby Moses and the baby Jesus, the Messiah was born into a world arraigned against him. After overcoming almost impossible odds, he will triumph: "The glory of the LORD will be revealed, and all flesh shall see it together, for the mouth of the LORD hath spoken it" (Isaiah 40:5).

Like David Lowenthal, Herbert Butterfield (1900–1979), another historian, criticizes what he termed "The Whig Interpretation of History."[46] Butterfield notes that in these accounts it proves impossible for the narrator to not look back upon historical events and trace from them a continuous line of "development" that ends in us being the heirs of the great tradition. This deep wish is made fully explicit in Christian readings of Jewish history, for the latter is but prelude to final revelation made clear in the birth of Jesus. The same wish is less blatant in Macaulay's history. Butterfield criticizes especially historians of science—and scientists—who looking backward to "great" names imbue them with modern sensibilities and concerns, overlooking the issues and concepts that animated these historical figures. Thus, some historians of science present Isaac Newton as a "pure" scientist while they dismiss or gloss over his lifelong fascination with mystical thought.

In sharp contrast, Butterfield emphasizes the integrity and distinctiveness of the past. The historian ought to seek ways to understand a worldview and consciousness different from ours. His (or her) effort should be "to understand the past for the sake of the past, and though it is true that he can never entirely abstract himself from his own age, it is none the less certain that this consciousness of his purpose is very different one from that of the Whig historian, who tells himself that he is studying the past for the sake of the present."

Butterfield's tone verges on scolding when he notes the seductive pull of looking at the past solely in terms of how its diverse actors and vastly complex questions pertain to our sense of who we are. In other words, historians can fall into the trap of seeking heritage. They will always err by either omitting the facts and interpretations that do not fit their story or by imposing on the past an order and meaning derived from their emotional needs: "The historian is bound to construe his function as demanding him to be vigilant for likenesses between past and present, instead of being vigilant for unlikeness; so that he will find it easy to say that he has seen the present in the past, he will imagine that he has discovered a 'root' or an 'anticipation' of the twentieth century, when in reality he is in a world of different connotations altogether, and he has merely tumbled upon what could be shown to be a misleading analogy."

Butterfield makes an eloquent plea against the dramatization of history: "Instead of seeing the modern world emerge as the victory of the children of light over the children of darkness in any generation, it is at least better to see it emerge as the result of a clash of wills, a result which often neither party wanted or even dreamed of, a result which indeed in some cases both parties would equally have hated, but a result for the achievement of which the existence of both and the clash of both were necessary."

Yet, why is analogous thinking so attractive? If we exclude from consideration the most extreme examples of chauvinism and special pleading, do we wish to exclude all reflection upon American heritage, for example? Butterfield's admonitions are fitting and appropriate reminders to academic historians that their task is not to champion their particular nation but to elucidate the many ways in which we do not know the past. Historians ought to be sober spoilsports who question deeply any hint of theological intent in themselves or others since that intent will shape how they examine the past. This means that a sober history cannot generate the same kind of dramatic tension, plot, and sureness of providence evident in popular histories.

As Lowenthal says, "History and heritage alike apprehend the past. But what they find and transmit, and why, are quite different. History tells all who listen what supposedly happened, suggesting how things came to be as they are. Heritage passes on exclusive myths of origin and *continuity*, endowing a select group with power and prestige."[47] This is too cynical if Lowenthal means that all talk about heritage is that of self-interested groups who perpetuate themselves, like royal lines in an ancient duchy of the court of George III.

Butterfield's admonitions are well meant and well said: yet can we escape this kind of discourse? To confront slavery, what discourses were best? As I noted earlier, history shows neither Britain nor the United States ending the slave trade except at the insistence of religious zealots. The Quakers insisted that sharing a common Father meant that all humans were essentially of the same family. Just as I could not morally enslave a brother or sister, I could not enslave an African. None of these claims arise from empirical study or historical scholarship. The struggle between North and South in the American Civil War was, in this sense, a religious struggle, a struggle of humanistic thought, about how we should demarcate the boundary of "full-fledged human being." Thus, we also fall into the gap of not knowing. Like Percival Lowell and J. B. Rhine, we struggle to discern in the noisy matrix portents of what we should do and who we might become. Texts and methods that we label "great" are those that supply an answer. For Americans, among such texts are the statements of the founders and our most esteemed presidents.[48] For Westerners, the efforts of two great twentieth-century European thinkers, Sigmund Freud and Ludwig Wittgenstein, provided new methods—and therefore new hopes—to demarcate hidden truths, as we will see in the next chapter.

6. SEARCHING FOR ESSENCES:
FREUD AND WITTGENSTEIN

SEEING INTO THE PSYCHE: FREUD'S DIAGRAMS

Sigmund Freud, still the most cited person in psychology, began as a scientist looking intensely at the nervous system. The need to earn a living to support his family, and perhaps Viennese anti-Semitism, drove him out of the research university and into private medical practice. There, treating neurotic patients, he developed the discipline that was to become psychoanalysis. In doing so, Freud moved from the study of a natural object, the nervous system organized like all natural objects into hierarchies, to the study of the brain and the mind, the latter a natural object and a cultural object. To the degree that the mind is "cultural," it is organized horizontally, by associative networks. Deciding how to demarcate brain (the natural object) from mind (a cultural object) is an as-yet-unmet challenge.

While cultural objects—such as personal and public narratives—are as real and as important as the brain, they will not yield to microscopic study; they will not give up their secrets by magnifying them into ever-larger images. On the contrary, when we magnify cultural objects, including representations of the mind, we generate noise, not signal. Where there is only noise, the only source of organized structure is what we impose upon the noisy array. This makes humanistic theories and schools prone to paranoidlike reasoning. It also helps explain why religious sects and other groups dependent upon esoteric teachings are prone to politicization: without external criteria with which to adjudicate disputes, we rely upon force and political persuasion.

Freud trained in the rigorous schools of neurological research conducted in Vienna under the banner of Hermann von Helmholtz, an acknowledged medical genius of the nineteenth century. Among Helmholtz's followers were Ernst Brücke, Freud's first instructor in neurology. Charles Darwin and other luminaries cited Helmholtz regularly.[1]

Like Helmholtz and his illustrious students, Freud struggled to explain how vision and other sensory processes occurred within the brain, how vision differs from hallucination, and how we can learn to distinguish between the two. As we have seen, the earliest Greek philosophers asked the same question: How can we distinguish mere opinions, which can be wrong, from true knowledge, especially from timeless truths like that afforded by mathematics? The latter seemed to emerge from concentrated thought, from looking inward and examining the objects of mathematics and geometry arranged within the ideal space of pure thought.

Agreeing with this notion, René Descartes, a brilliant seventeenth-century mathematician and methodologist, made deductive reasoning paramount to all forms of thinking. With that idealization came the heavy hand of Cartesian rationalism and its metaphysical assumption that the human mind differs from the human body, the latter separate and inferior to the former. In addition to Descartes's influence, common sense tells us that we can examine our minds in ways that we cannot examine the natural world. Finally, added to this puzzle was the deeper mystery of human consciousness.

Given the vastness of the problem of consciousness, it seems prudent to put it aside and concentrate instead upon problems that lend themselves to scientific investigation. This brings us back to neurology and the study of vision. In this section we consider Freud's neurological education and his hope that his new discipline, psychoanalysis, would become a mode of investigation analogous to that of other observational disciplines. I will suggest that this is partly correct and partly wrong. It collapses the search for patterns of a patient's behaviors into another kind of search, the examination of the inner world, by breaking it into parts, by analyzing it.

Freud led psychoanalysis into a quest for essences, that is, properties of the mind (or the instincts or the drives) whose nature we can deduce by examining fantasies, wishes, and the multitude of human behaviors that come under the purview of the clinician. As Alva Noë and Evan Thompson say in their preface to an anthology of readings on vision and philosophy, it would be difficult to overstate the degree to which these problems are alive and unresolved.[2]

Dedicating himself to exact observation in the most rigorous laboratories of his day, Freud used microscopes and enhanced staining techniques to see and

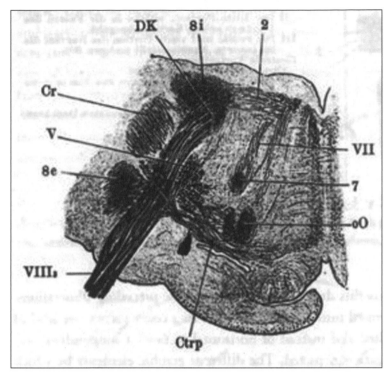

Figure 6.1. Freud's diagram of auditory nerves.

then to render the images that emerged. By 1879, Freud had published several studies on the use of gold chloride to study tissues from the medulla oblongata.[3] Evaluating a neuroanatomical drawing by Freud from 1886, Cornelius Borck notes, "What strikes us once again are the precise details of the illustration."[4] From these exact, almost photographic renditions of tissue structure, Freud developed diagrams of nerve functioning. While these diagrams abstract key elements from the anatomical descriptions, they refer closely to the original visualizations.[5]

The dark bundle labeled "VIII" is the auditory nerve; it enters into the oblongata, then branches to the upper region (8i). From there, a part of the auditory nerve leads to the center of the tissue (Ctrp). As Borck notes, Freud's diagram shows precisely how the nerve fibers cross one another. To put this another way, Freud's diagram is a strong model of a natural object. By studying this diagram, another trained neurologist could deduce important truths about the auditory nerve. With better magnification we learn even more. Some 120 years after these initial studies, anatomists and others can show with vastly greater detail how the auditory nerve functions.

Everything changes in Freud's monograph *On Aphasia*. Because he did not accept the claim that speech originated in a fixed, specific region of the brain,

Freud did not localize speech functions on an anatomical map. As Freud says, "The speech apparatus as conceived by us, had no afferent or efferent pathways of its own."[6] Instead, Freud argued against the localization hypothesis and used diagrams or schemata of speech functions to support his point. These schemata depict plausible lines of connection between functional centers, such as "word representations" and "object presentations." Given this functional description, Freud's diagrams have no direct relationship to brain structures or brain locations. When Freud wished to illustrate how the nerve cells *might* communicate with one another he used a sketch of how these "relays" operated.

Freud's decision to abandon direct, anatomical descriptions becomes abundantly clear in "A Project for a Scientific Psychology" of 1895.[7] In this unpublished tour de force, Freud sets out a series of brilliant hypotheses about how different hypothetical groups of nerve cells give rise to psychological phenomena. In the project and in the progeny to which it gave rise, Freud attempts to marry neurological discourse to descriptions of psychological functioning. By the time he published his book on dreams, Freud ceased to use graphical illustrations that had any relationship whatsoever to anatomy. On the contrary, the famous illustrations of "regression in dreams,"[8] for example, merely illustrate the concepts Freud announces in the text. Each is a "general schematic picture of the psychical apparatus."[9] As *illustrations* of concepts they are no longer renditions of natural objects; we cannot study them to gain new information about brain tissues, for example.

Thus, speaking of "neurones" or nerve cells, Freud talks about the Ψ-system, which he later will call the ego: "A most promising light would be thrown upon the conditions governing the excitation of neurones if it could be confirmed that *in the Ψ-systems memory and the quality that characterises consciousness are mutually exclusive.*"[10] Many smart people have spent hundreds of hours teasing out precisely what Freud meant.

Freud's diagram of the ego and the id illustrates his conception of how the id, that part of the brain-mind hypothetical structure, is linked to repressed mental contents.[11]

Information flows from the external world through the *Wahrnehmung-Bewusstein* (W-Bw), the "perceptual consciousness," to the *Vorbewusstein* (Vbw), preconscious (Pcs.) to the ego. Because the "repressed is cut off sharply from the ego by the resistance of repression," it can communicate with the ego (*Ich*) only through the id (*Es*). The diagram represents this concept by the graphic slice between the ego and the repressed (*Verdrangung* or Vdgt.). There is, of course, Freud notes, no such slice in neural tissues in which repression and other psychological events occur. (Like the equator, these are imaginary lines.) Freud returns to his earlier studies of acoustic anatomy when he sticks a "cap of hearing" (labeled "*akust.*") on top of the ego.

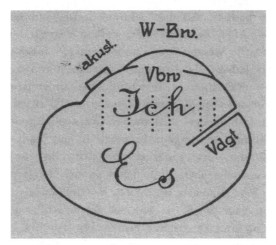

Figure 6.2. Freud's illustration of ego and id.

When he changed his mind about these theoretical matters, Freud changed these diagrams. He recast these schemata in 1925 and again in his essays on ego psychology.[12] In these latter texts the terms *ego, id,* and *superego* name psychological functions that Freud ascribes to hypothetical mental structures. These diagrams have no direct relationship to the brain or any other nerve tissue. Thus, depending upon his sense of the importance of the id concept, Freud draws it to different scale; larger in 1923, smaller in 1932.[13]

Drawings of concepts are not maps; we cannot examine them and deduce from them anything true about nature. In illustration 6.1, one can examine Freud's drawings and learn how one part of the cell, for example, pertains to another part. An accurate map of Manhattan tells us that Central Park is "above" Wall Street because the map shows us how the two areas pertain to one another. In illustration 6.2, of the ego and the id, we can learn nothing about the brain (or the mind) since it does not portray a thing, but a concept.

Freud was keenly aware of this difference. It amounted to sea change—he had begun as a natural scientist, now he became an advocate of metaphorical reasoning. The leap from neurology to metaphor happens openly in *The Interpretation of Dreams*. There Freud brings us back to the world of microscopes and science:

> Everything that can be an object of our internal perception is virtual, like the image produced in a telescope by the passage of light rays. But we are justified in assuming the existence of the systems (which are not in any way psychical entities themselves and can never be accessible to our psychical perception) like the lenses of the telescope, which cast the image. (Standard Edition 5.611)[14]

Freud recognized that his new science, psychoanalysis, was not like his old science, neurology. The latter focused upon natural objects that one could investigate objectively and in increasingly deep ways, namely tissues and nerve cells. Because research scientists could see these new objects, made visible after excruciating work, they could draw or photograph them and make those renditions available for public scrutiny. As images of natural objects, these drawings and photographs could be studied, and new truths deduced from them.

Freud's new science (or new discipline) focused upon objects that were *not* visible. Indeed, he says that the objects that psychoanalysts examine, such as dreams, wishes, fantasies, and thoughts, are "virtual objects" in the way that images of a planet produced by a telescope are "virtual." By examining these virtual images, we can learn much about the planet, and we can learn much, but not everything, about the lenses that bring the planet's image into focus. The ego and its various functions are like these lenses: we learn about them indirectly, by examining their effects.

This is a brilliant metaphor. Is it also illuminating? Do we agree that the thoughts and fantasies are like the virtual *images* in telescopes and microscopes? A great deal hinges on our answer. If we agree with Freud that thoughts and fantasies are like virtual images, then the metaphor becomes a strong model of the psyche. Yet, we note that this metaphor-model of psychic functioning begs the question: Are virtual images real things visible to any observer? Experimenters can fit a telescope or a microscope with multiple viewing devices so that three people can observe the same virtual image at the same time. The problem with thoughts and fantasies is that we don't yet know what they are. They are *like* virtual images, true, but they are also *like* photographs or mirages. Thoughts are also unlike either of these. We cannot observe either our own thoughts or those of another person the way we can observe a virtual image.

Why does Freud offer this comparison? One reason is that the camera and other imaging mechanisms seemed to offer new models of human perception.[15] Freud retains this model of virtual images, I suspect, because it gives psychoanalysis an observational basis. Thus, Freud introduced medical students to psychoanalysis in 1915–1916:

> In medical training you are accustomed to see things. You see an anatomical preparation, the precipitate of a chemical reaction, the shortening of a muscle as a result of the stimulation of its nerves. Later on, patients are demonstrated before your senses. . . . Thus a medical teacher plays the main part of a leader and interpreter who accompanies you through a museum, while you gain a direct contact with the objects exhibited and feel yourselves convinced of the existence of the new facts through your own perception. In psycho-analysis, alas, everything is different. (Standard Edition 15.16–17)

Medical students learn about diseases by seeing their effects in patients or in tissues or in cells. Seeing things is the preferred route of true knowledge. According to the Greeks, it offers us the most exact way to distinguish between similar objects. Plato says that vision is "the most sun-like of all the instruments of sense" (*Republic* 508b).[16]

Seeing things is preferable; but, alas, the objects of psychoanalysis are internal and virtual. They cannot be demonstrated to a skeptical audience that demands the usual form of proof. By locating the objects of his science within the mind, Freud renders them invisible; we can know them only indirectly. In making this claim, Freud knows that he teeters on the edge of Romanticism. This echoes German Romantics, like the poet Novalis (1772–1801), who said, "Every procedure of our mind that is made conscious is, in the strictest sense, a newly-discovered world."[17] While this matches Freud's claim to have discovered new realms, to have become the conquistador of the unconscious, he wished to remain grounded upon the science of his day, the world of neurology and other observational disciplines.

Before his audience rises to a skeptical rejection, Freud asks them to attend to their dreams, study their parapraxes, and observe their own mixed feelings and other neurotic actions. All these together will suggest—if not demonstrate—the reality of the objects of psychoanalytic inquiry.[18] Yet, such investigations cannot answer the basic question: Are ideas, affects, and fantasies parts of the natural (that is, neurological) world? On the contrary, to study parapraxes and dreams is to study complex behaviors that may or may not correlate with self-reports and patterns. This is the stuff of clinical encounters, but it's not a study of natural objects.

By affirming the centrality of vision, Freud follows the dictates of the Greek authors who, as we saw above, made vision preeminent. It was the conduit of reason, a source of true knowledge. However, the Greeks also distinguished between merely seeing the world of change and really seeing, that is grasping intellectually, the underlying, invisible, world of permanence and truth. To love the activity of seeing is a form of doxophilia; as Plato says, it is to love opinions and sensations: "Do we not remember that we said that those loved and regarded tones and beautiful colours and the like, but they could not endure the notion of the reality of the beautiful itself?" (*Republic* 480a). We prefer vision, Aristotle says in the famous opening paragraph of *Metaphysics*, because "of all the senses sight best helps us to know things, and reveals many distinctions."

However, even vision is limited; like all senses, it can falter. Even when accurate, vision helps us discern only transitory events. It cannot tell us about the unseen, permanent, reality that underlies transition and change. Aristotle says, "Of the qualities there described the knowledge of everything must necessarily belong to him who in the highest degree possesses knowledge of the universal, because he knows in a sense all the particulars which it comprises. These things, viz. the most universal, are perhaps the hardest for man to grasp, because they are furthest removed from the senses" (*Metaphysics* 980a).[19] A chief source of this metaphysics was the stunning success of Greek mathematics, especially the use of the deductive method to discover the existence of π and other transcendental entities.

Advancing a similar claim for the objects of structural anthropology, Claude Lévi-Strauss held that the three great reductive disciplines of the nineteenth century—psychoanalysis, Marxism, and geology—proved that valid understanding "consists in the reduction of one type of reality to another; that true reality is never the most obvious of realities, and that its nature is already apparent in the care which it takes to evade our detection."[20] In brief, thanks to the Greeks, Western authors presume that there are two worlds, the world of change and disorganization, and the world of unchanging reality that lies behind the flux. This is in striking contrast with other systems, such as Chinese philosophy which assumes that one world is "the source and locus of all of our experience."[21]

Greek mathematicians had, through deductive reasoning, discovered the nature of these hidden entities.[22] Although Freud did not claim strict logical rigor in inferring the likely structure of the ego's lenses, he does presume that his clinical descriptions will eventually be matched against empirical discoveries about the neural substrate. Like the Greeks and their nineteenth-century German admirers, Freud holds that only science can claim knowledge of the underlying and permanent reality. From his earliest essays on neurology in the 1890s, to his last major publication, in 1940, Freud affirmed this claim: "Behind the attributes (qualities) of the object under examination which are presented directly to our perception, we have to discover something else which is more independent of the particular receptive capacity of our sense organs and which approximates more closely to what may be supposed to be the real state of affairs. We have no hope of being able to reach the latter itself, since it is evident that everything new that we have inferred must nevertheless be translated back into the language of our perceptions."[23]

The cleverness, yet epistemological shakiness, of Freud's notion of "virtual objects" has long been recognized. Analysts like Charles Brenner and Jacob Arlow, who assert that psychoanalysis is among the observational sciences, retain Freud's metaphor and claim it has explanatory value. They say, "Theories of astronomy based on data obtained without the use of a telescope would be very different from those based on data obtained by using a telescope, photography, and spectroscopy. Similarly, theories of infectious disease will vary depending upon whether one uses a microscope or not. The same is true for theories of the mind. The data available by means of introspection and commonplace observation of the behavior of others, for example, are very different from the data available when one uses the psychoanalytic method in a psychoanalytic situation."[24]

Psychoanalysts who reject Freud's metaphor of virtual objects also reject the claim that psychoanalysis is among the observational sciences. Michael Basch systematically deconstructs Freud's earliest comments on the "virtual" objects metaphor, in "A Project for a Scientific Psychology." And Donald Spence argues that psychoanalysis still lacks definite rules for bridging metaphors like this with specific observations.[25] Among Spence's trenchant criticisms of traditional psychoanalytic epistemology is his comment that analysts love metaphors of the mind, but offer few ways to connect one metaphor with another. That is, while we admire Freud's brilliance, in likening thoughts to virtual images, we cannot align this metaphor with any anatomical structure, nor with any observable behaviors. It does not rise to the level of model of the mind.

We cannot, therefore, agree with Charles Brenner, who writes: "Just as Galileo learned previously unsuspected facts about the moons of Jupiter simply by looking through his newly invented telescope, just as Pasteur revealed a whole new area of biology by applying to it the use of the compound microscope, so Freud, by devising and developing the psychoanalytic method, made possible the most important discoveries in the field of human psychology in the recorded history of science."[26] Metaphors are evocative, they expand the range of possible connections, but they cannot serve as demarcations.[27] How to justify the turn to metaphor is an ancient problem of psychoanalytic theorists.[28] We return to telescopes and microscopes in the next chapter.

WITTGENSTEIN AND SHARP FOCUSING

A similar wish to isolate essential truths animated Ludwig Wittgenstein's first book, *Tractatus Logico-Philosophicus* (1922).[29] The young Wittgenstein believed that by magnifying an idea or the meaning of a sentence, we can enlarge it and discover its truth. Throughout the *Tractatus* Wittgenstein says that he is searching for crystalline truths, the atomic facts, upon which he could build a rigorous philosophy. Impressed by the immense advances made in the natural sciences by the 1920s, especially the Einsteinian and Quantum Theory revolutions, Wittgenstein and others renewed the effort to secure a similar foundation for philosophy. A common diagnosis offered of the failure of philosophy to keep pace with the advance of science turned on the analysis of language and philosophic pretensions to special knowledge.

In the *Tractatus*, Wittgenstein rejects a constructive role for philosophy. It will not join the natural sciences and therefore cannot progress as they progress. Philosophy will not add new truths to our description of the real world (this is the task of science). On the contrary, philosophy delimits the range of knowable and useful propositions and distinguishes them from the unknowable and useless. Wittgenstein arranged his comments in numbered sections, the following are from #4:

4.11   The totality of true propositions is the total natural science (or the totality of the natural sciences).

4.111   Philosophy is not one of the natural sciences.
(The word "philosophy" must mean something which stands above or below, but not beside the natural sciences.)

4.112   The object of philosophy is the logical clarification of thoughts.
Philosophy is not a theory but an activity.
A philosophical work consists essentially of elucidations.
The result of philosophy is not a number of "philosophical propositions," but to make propositions clear.
Philosophy should make clear and delimit sharply the thoughts which otherwise are, as it were, opaque and blurred.

4.113   Philosophy limits the disputable sphere of natural science.

4.114   It should limit the thinkable and thereby the unthinkable. It should limit the unthinkable from within through the thinkable.[30]

To do this, to delimit the thinkable, Wittgenstein and others searched for a formal language, shorn of poetic shadows and endless, associative linkages. The rigor of Truth Tables (inherited from Gottlob Frege and others) provided him a graphical representation of what "valid" and "true" meant. It also allowed him to deduce that contradictions and tautologies say nothing about "this world" since there are no conditions under which the first might be true or the second might be false:

4.461   The proposition shows what it says, the tautology and the contradiction that they say nothing.
The tautology has no truth-conditions, for it is unconditionally true; and the contradiction is on no condition true.
Tautology and contradiction are without sense.
(Like the point from which two arrows go out in opposite directions.)
(I know, e.g. nothing about the weather, when I know that it rains or does not rain.)

4.462   Tautology and contradiction are not pictures of the reality. They present no possible state of affairs. For the one allows *every* possible state of affairs, the other *none*.

When used correctly, propositions in the natural sciences admit of being true or false under certain conditions:

4.463   The truth-conditions determine the range, which is left to the facts by the proposition.
(The proposition, the picture, the model, are in a negative sense like a solid body, which restricts the free movement of another: in a positive sense, like the space limited by solid substance, in which a body may be placed.)
Tautology leaves to reality the whole infinite logical space; contradiction fills the whole logical space and leaves no point to reality. Neither of them, therefore, can in any way determine reality.

4.464   The truth of tautology is certain, of propositions possible, of contradiction impossible.[31]

Wittgenstein is not suggesting that tautologies and contradictions are useless junk to be abandoned. When he says that "tautology and contradiction are without sense," he means that they are not sentences that we can map onto any part of the real world. Rather, he says that they define the limits of propositions: "4.4611. Tautologies and contradictions are not, however, nonsensical. They are part of the symbolism, just as 'o' is part of the symbolism of arithmetic."[32] To assume that they must also be names of a real place, a realm beyond this world, is to engage in metaphysical musings. Metaphysical urges to go beyond the propositions of the natural sciences should be resisted. In the book's famous last sentence, Wittgenstein insists "Whereof one cannot speak, thereof one must be silent."[33]

Another way to explain the success of the natural sciences was to note that natural scientists focused their attention upon smaller and smaller parts of nature, upon atomic and generalizable facts. In order for this to occur, two features are relevant. First, nature is organized in discrete layers of complexity, and with the proper attitude and using the proper magnification, we can go from higher to lower levels. With more magnification comes more information, and with more information comes more insight. Second, these discrete levels are relatively independent of one another. We can investigate higher-level structures even if we don't know much about the levels below them. Nature seems capable of almost infinite reduction, from layer to layer, from macro, to meso, to micro, and by the early twentieth century, to subatomic levels. At each level we find new information and richer sources of data. Unless we're hovering at the limits of resolution, there is *always* more to see. When they reach the limit of resolution, as we shall see, sometimes scientists act just like humanists.

This is precisely not true of cultural artifacts, such as news photographs or dreams of patients in psychoanalysis or poetic and religious texts. These artifacts are not organized such that we can examine them using ever increased magnification. Nor are they organized in obvious, discrete layers of complexity. In *Philosophical Investigations* (1953) Wittgenstein offers a self-analysis of his youthful zeal to grasp the essence of a sentence or an idea. There are many ways to read Wittgenstein and many ways to parse his arguments in both of his major texts. I emphasize his psychological insight into the wishes that animated the writing of the *Tractatus*.[34] I then stress his later psychological insight that we wish to know where everything sits and that this metaphysical wish cannot be granted through philosophy.

MAGNIFYING TRUTHS
IN PHILOSOPHICAL INVESTIGATIONS

Having set out a deductive theory of language and logic in the *Tractatus*, in *Philosophical Investigations* Wittgenstein undoes it. Chief among his targets is the yearning for metaphysical certainty, the search for the essence of language to be achieved by peering at it with special intensity.[35] Sharing Freud's wish to match the Greek discovery of transcendental truths, such as the nature of π, the young Wittgenstein wrote the *Tractatus* as the last word on problems of language. Thirty years later he offered a psychological reading of his younger self and his youthful wishes:

> 36. Where our language suggests a body and there is none: there, we should like to say, is a *spirit* [*ein Geist*]."

> 92. This [seeking of exactness] finds expression in questions as to the *essence* of language, of propositions, of thought.—For if we too in these investigations are trying to understand the essence of language—its function, its structure,—yet *this* is not what those questions have in view. For they see in the essence, not something that already lies open to view and that becomes surveyable by a rearrangement, but something that lies beneath the surface. Something that lies within, which we see when we look into the thing, and which an analysis digs out.
> "*The essence is hidden from us*": this is the form our problem now assumes. We ask: "*What is* language?," "*What is* a proposition?" And the answer to these questions is to be given once for all; and independently of any future experience.[36]

The philosophic analyst digs out hidden meanings or, as Freud often said, the psychoanalyst acts like an archaeologist who digs into the layers of history and unconsciousness. In both cases, the analyst penetrates the mere surface of things, which anyone can observe, and peers beneath. Nature or the Truth hides from us. We intrepid investigators must pursue her. What gives us this superior vision? It is not better eyesight or a better instrument, say, an improved telescope. Anyone can use a telescope to see the surface of things once hidden from sight. If everyone can use a telescope to see the surface of the Moon, it is no longer hidden. School kids with sticky fingers can see the Moon, and that means that discourse about the Moon is no longer interesting as the subject of reflection.

No; our superior vision derives from the spiritual strength we gained by intense immersion in rigorous reflection not available to the public. This rare discipline gives us a *new* instrument. This new instrument is rigorous thought:

> 97. Thought is surrounded by a halo.—Its essence, logic, presents an order, in fact the a priori order of the world: that is, the order of *possibilities*, which must be common to both world and thought. But this order, it seems, must be *utterly simple*. It is *prior* to all experience, must run through all experience; no empirical cloudiness or uncertainty can be allowed to affect it.—It must rather be of the purest crystal. But this crystal does not appear as an abstraction; but as something concrete, indeed, as the most concrete—as it were the *hardest* thing there is.[37]

This kind of thought is intense reflection done by a small group of elites. It has a halo because it traffics in transcendental truths the way that angels and saints, shown with halos, personified otherworldly truths in the Christian tradition. The goal of these thinkers is to name the utterly simple, that is, the fundamental and eternal truths that precede all possible experience. For everyday experiences are superficial, expressions of deeper structures, or the Structure, which shapes them. In the young Wittgenstein's phrase, these truths or this Structure is like a crystal, like a diamond, the hardest of all natural things, simple, pure, and eternal.

The pursuit of this eternal truth is a religious sentiment. It is the wish to see deeper, to magnify what seems invisible. Like the great mystics, the true thinker seeks moments of uncanny recognition: to see the world's hidden patterns, to identify the source of these patterns, the Patterner. To get closer to the latter we must search for deeper or larger or higher concepts:

> 97. We are under the illusion that what is peculiar, profound, essential, in our investigation, resides in its trying to grasp the incomparable essence of language. That is, the order existing between the concepts of proposition, word, proof, truth, experience, and so on. This order is a *super*-order between—so to speak—*super*-concepts.

> Whereas, of course, if the words "language," "experience," "world," have a use, it must be as humble a one as that of the words "table," "lamp," "door."[38]

Having offered himself as a case study of this kind of yearning, Wittgenstein presents a rebuttal. The word *language* has no more status than homely words *table* and *door*. As an antidote to transcendental longing, Wittgenstein asks us to observe actual language, used in actual circumstances. Employing terms from this book, I read this as an admonition to refrain from projecting onto our subject matter a demand for depth because that demand takes us away from observation. By beginning with the demand for depth we impose an ideological screed on everything that we see:

> 107. The more narrowly we examine actual language, the sharper becomes the conflict between it and our requirement. (For the crystalline purity of logic was, of course, not a *result of investigation*: it was a requirement.) The conflict becomes intolerable; the requirement is now in danger of becoming empty.[39]

There is a kind of pleasure in searching for the deeper Truth because it occurs in a world where mere observations do not impede us. The conditions are ideal because we feel free to "think"—that is, to speculate—without the constrictions of uncertainty and the burden of observation. Wittgenstein's radical criticism of himself extends to Freud and Freudians who believe that by intense training in learning a special "gaze" one can grasp something true about the inner world of our patients. In an even more radical criticism, Wittgenstein asks us to consider whether that notion, championed by Descartes and metabolized into ordinary discourse about the inner world and the "self," is useful.

THE MAGNIFICATION FANTASY
AND IDEOLOGICAL LEANINGS

Freud and Wittgenstein sought to make their disciplines adhere to the rules of natural science, as each understood it in the early twentieth century.[40] This is not surprising for it locates each discipline within the realm of an advancing research paradigm. Freud's insistence that psychoanalysis was a science, or if not yet a science, would become a science, was essential to his self-understanding. We have seen that when he left neurological imaging and took up the study of fantasy and metaphor, Freud abandoned all possibility of finding essences—at least in the ways typical of standard investigation.

Yet, he retained the language and value system of magnification, including the ideal of introspection and depth. This leap from neurology to psychoanalysis produced one of the great revolutions in Western thought; it made possible self-critique and cultural critique in ways not otherwise possible. These and numerous other effects of the psychoanalytic revolution are important and worthy of study and respect.

One aspect of this revolution, long recognized by analysts themselves, was the odd ways in which psychoanalytic speculations became codified into theory, and theory became codified into Truth. As I note throughout this study, I find this one of the deeper puzzles of being a psychoanalyst, and I come back to it many times. Here, though, I go back to comment on the older Wittgenstein commenting upon his youthful wish to find the essence of language:

> 113. "But *this* is how it is ——" I say to myself over and over again. I feel as though, if only I could fix my gaze absolutely sharply on this fact, get it in focus, I must grasp the essence of the matter.[41]

The experience of compulsion, itself a problem that preoccupied Wittgenstein throughout his career, is more an emotional event than a logical task. We drive ourselves to find the hidden truths, to find the essences; we talk to ourselves again and again. We fall into perfectionistic beliefs and actions: "If only I could fix my gaze absolutely" describes an archaic wish that if we just find the proper stance toward a puzzle it will yield up its secrets to us. It is a wish that depends upon the fantasy that objects like "sentences" and "thoughts" will yield to magnification in the ways that natural objects yield up their secrets when examined with an absolutely sharp gaze.

Because Wittgenstein speaks so openly about his wishes and his yearning, it might be tempting to speculate about the underlying reasons—his particular story—that produced this manifest content. In other words, we might try to psychoanalyze Wittgenstein by using his writings, connected to his life-story, biographies, accounts, diaries, etc. If done with scholarship and care, this might be a fascinating exercise. It might yield a plausible story about Wittgenstein that connected some parts of his emotional and sexual life with some parts of his intellectual work. If that occurred, it would be an instance of psychoanalytic art, not of psychoanalytic science, as I argue next.

Freud asked his listeners to examine their own behaviors, their dreams, their slips-of-the-tongue, etc.; he did not ask them to gaze within themselves. On the contrary, he asked them to record and reflect upon patterns in their observable behaviors. If, for example, a young man finds that he cannot recall the name of a certain professor on one occasion, it's easy to believe that this is trivial, not a sign of emotional conflicts. To convince himself that something is amiss, he must record many such instances and find a pattern of forgetting, learning, self-admonition, followed by forgetting again. Contemporary psychotherapy of all types depends upon naming patterns of manifest behavior, then articulating their shared themes.

Contrary to the badge "depth psychology," Freud's strategy was horizontal, as it were, not vertical. He did not offer new ways to investigate the internal structure of mental contents. On the contrary, he extended and justified the notion of unconscious processes by examining numerous, analogous behaviors. Thus, in the twenty-four volumes of the Standard Edition, the largest collection of Freud's work, we find two volumes dedicated to dreams, one to jokes, three to religion (including his last book, *Moses and Monotheism*), one to anthropology, one to the psychopathology of everyday life, another to sociology, and numerous essays on Shakespeare, Greek plays, war, literature of all types, novelists, and grief. One might say that these are merely instances of applied psychoanalysis, that the core of Freudian discipline lies in the scientific papers. Yet, the care that Freud put into the applied essays exemplifies his strategy. To make the core claims of his new science plausible, he had to link them to insights and wisdom of the ancients, and to the texts of idealized moderns like Dostoevsky and Goethe—each a great humanist. More than that, Freud recognized that his new science could not sidestep issues traditionally left to the humanities and to religion. The latter had to be confronted, for it competed with the psychoanalytic movement: "Religion alone is to be taken seriously as an enemy."[42]

Humanistic literature of all types is crucial to Freud's explication of psychoanalysis both as illustrations of his clinical notions and as validations (of a sort). For example, in 1928 Freud contributed an essay on Dostoevsky to a German edition of the latter's celebrated novel *The Brothers Karamazov*, first published in Russian in 1880.[43] Reading Freud's essay is like taking apart a Russian stacking doll (a matryoshka). We read Freud's interpretation of Dostoevsky's interpretation of the chief prosecutor's interpretation of the mind of Dmitri Karamazov, who is accused of murder. Each of these is a fascinating and brilliant narrative; none is definitive or provable. For example, Ippolit Kirillovitch, the prosecutor, says about Dmitri: "I can picture the state of mind of the criminal hopelessly enslaved by these influences—first, the influence of drink, of noise and excitement, of the thud of the dance and the scream of the song, and of her, flushed with wine, singing and dancing and laughing to him! Secondly, the hope in the background that the fatal end might still be far off, that not till next morning, at least, they would come and take him."[44]

Kirillovitch's speech, the last he was to give, is "the chef-d'oeuvre of his whole life," as Dostoevsky says at the beginning of chapter 5. The speech consumes five chapters of the novel and demonstrates knowledge of Dmitri's mind so vast that only a great novelist (or great psychoanalyst) could have constructed it. Indeed, the speech is so brilliant that it persuades us, momentarily, that *this* is surely a valid and accurate portrait of the mind of a killer. Yet, because we are in the realm of persuasion, that is rhetoric, we recognize that eloquence and narrative power are merely charming; they do not demarcate the truth. Lawyer Fetyukovitch, Dmitri's defense counsel, promises to counter the prosecution: "His voice was a fine one, sonorous and sympathetic, and there was something genuine and simple in the very sound of it. But everyone realized at once that the speaker might suddenly rise to genuine pathos and 'pierce the heart with untold power.'"[45]

That we can find such striking parallels between literary examples and clinical ones does not make the latter nonsense. But it does suggest that in doing psychotherapy we focus upon horizontal events, a sequence of thoughts, patient reports, and therapist conjectures. We do not peer into the depths, vertically, searching with intensity upon a single event. When Freud asked us to examine our behaviors, dreams, and slips of the tongue he asks us to engage in a similar act of self-portraiture. Some are accurate, some are not, and some reveal painful truths about how we feel and imagine our worlds. Yet, none of them is akin to an X-ray or an electron microscope; none take us below the surface. In their best moments, such as *The Brothers Karamazov*, they take us across what seem impassible barriers.

CULTURAL ARTIFACTS AND REDUCTIONISM

Cultural artifacts cannot sustain reduction, at least using the simple model of scale and hierarchy. If we go beneath the surface of the painting, we lose the painting. The universal importance of music cannot be denied. But when we lose the performance of the whole, the music stops. Unlike ordinary intellectual work, musical performers can achieve greatness by repeating what has already been done. A young scientist will not win tenure in a university by reciting, repeating, or even teaching past scientific masterpieces. According to the rules of the university, tenure is awarded to those who contribute *new* knowledge to their disciplines. What do musical performers contribute that is novel to their discipline? For example, what value does B. B. King add to a few dozen traditional songs that, when reduced to statements, tell us nothing new? Consider the chorus of "The Thrill Is Gone":

The thrill is gone
the thrill is gone away.

I know the thrill is gone
the thrill is gone away.
Now you've done me wrong
you'll be sorry someday.[46]

Removing the redundancies, we can summarize: "You hurt me, I'm sad, you'll be sorry." From twenty-nine words to eight, a savings of 72 percent.

What's wrong with this form of reductionism? The traditional answer is that we value new performances because they give us access to feelings that sustain us and to which we have no other route. When I hear B. B. King sing "The Thrill Is Gone," I am changed.[47] My emotional life deepens; I discover a new way to feel longing, loss, and reconstitution through this moment of art. I am grateful because he has given me a way back to part of my life, the parts about longing and boyhood that I cannot recall any other way—except through art. If a musicologist stops my CD and gives me a detailed history of how blues chords developed, how this part of the song differs from that part, I will learn something—but the thrill is gone, at least for me.

To reduce a cultural artifact, like a poem, into a message, into a formal communication, destroys the poem. We come back to Wallace Stevens's poem "Anecdote of the Jar" (1919):

I placed a jar in Tennessee,
And round it was, upon a hill.
It made the slovenly wilderness
Surround that hill.
The wilderness rose up to it,
And sprawled around, no longer wild.
The jar was round upon the ground
And tall and of a port in air.
It took dominion every where.
The jar was gray and bare.
It did not give of bird or bush,
Like nothing else in Tennessee.

It would be grotesque to summarize the poem in a headline: "Man places jar on ground, observes that jars don't grow on trees." With wonderful compression Stevens transforms the jar into a centralizing presence that shakes nature into an order that no longer sprawls. Nature is subordinated: it pays obeisance. Responding to this poem, critics draw linkages to historical struggles (Whites against Indians), past poems by past great poets, political and artistic movements, and irony. One critic says, "The poem is, of course, an ironic critique of the Romantic yearning for the creative interfusion of consciousness and nature as the basis of art."[48] Some critics argue that Stevens is looking back to Keats's "Ode on a Grecian Urn" (1819) and to Romantics not relevant to the modern period. Others focus upon Stevens's contemporaries, like Marcel Duchamp, an ironic, visual artist who made art out of everyday manufactured objects, like urinals: "It is not the execution but the idea behind the work that makes the readymade interesting."[49]

Yet other critics place the poem in a larger discussion of literary theory. For example, formalist critics emphasize the structure and pattern of poetic expressions and put less emphasis upon idiosyncratic contents and sounds.[50] B. J. Leggett focuses upon a word: "Just as the sound of *round* appears to dominate the poem once we fix on it—so too does our understanding that the jar orders the wilderness depend on our adopting momentarily the point of view of the jar." Leggett then ascribes a general theory to Stevens: "And in the end, by asserting the jar's alien presence in a readily apprehended world of bird and bush, the instigator of this epistemological exercise unmasks a naive empiricism that lies behind the more official perspectivist stance of the poem."[51]

How would we prove Leggett wrong? We could ask Wallace Stevens, via séance. Yet even if he were available and willing to talk with us, Steven's reading of his poem is just another interpretation. Once it has become a work of art, the poem ceases to be a translatable utterance. It is not a message that can be reduced into simpler terms, nor into the artist's intentions. To support one reading over another we must generate a larger account of the poem and show how other critics got it wrong. Because authoritative readings of past works enter into discourse about the poem, newer and purportedly better readings must wrestle with these past readings.

These subtle rereadings will persist as long as the poem persists. Like religious texts, poetic masterpieces do not let us claim that we have exhausted their meanings. In this important way, they and discourse about them differ from that of level I and level II. To the degree that language about the self belongs to this domain, it cannot be reduced to language about the ego or language about the brain.

I do not dismiss the urge to find essences and to find the hidden truth of things. People who have sought for essences are among the most creative and thoughtful in Western history. Like the scene in *High Anxiety*, we seek to find the truth by magnifying what we believe we know. Like the young Wittgenstein of the *Tractatus*, we seek for essences, the interior structure of what appears on the surface. We wish to go beneath it and discover what we cannot see. When we pursue the same program with cultural artifacts, with nonnatural products shaped to communicate information, trouble emerges. For these items—the camera film, the newsprint, the surface of a painting, a novel, and the infinite variety of signs that flood over us daily—do *not* yield to magnification.

This brings us back to our original issue of seeing and magnification. Galileo used telescopes to see entirely new things in the early seventeenth century. Jan Swammerdam and Anton van Leeuwenhoek used microscopes to see new things in the late seventeenth century. Around 1662, Swammerdam refuted Descartes's notion that muscular contraction was caused by the brain's "animal spirits" when he showed that one could cause a frog's leg muscle to contract by irritating the nerve fibers attached to it.[52] This fundamental experiment, repeated by many millions of high school biology students, helped spread biological reductionism throughout the scientific world.

Using improved optical instruments, European and American astronomers saw new things in the late nineteenth century. Like their Dutch counterparts, nineteenth-century astronomers applied old categories to new data.[53] When they reached the limits of detection they reached the limits of resolution, and some astronomers went beyond the information (or signal) available to them. In going beyond the signal they stumbled upon randomness (or noise). Upon that noise they projected their wishful hopes. Wishing desperately to find evidence of extraterrestrial intelligence, as we saw in chapter 5, Percival Lowell asserted that the vague lines he saw under low magnification must be "canals," vast water-carrying devices erected by Martians. So sure was he of this truth, that Lowell wrote three books on Mars—the first on his telescopic findings, the second and third on Martian sociology.[54]

This question is especially important to psychoanalysts because our discipline requires us to interpret the meaning of patients' thoughts, actions, and dreams. It is easy to see that some dreams are meaningful. Consider a dream in which a young male academic clobbers the older male analyst with a dictionary (having been humiliated in the previous day's session when the analyst corrected his grammar). This dream means something obvious: that the patient was humiliated by the correction, felt enraged but powerless to revenge himself, and later that night dreamt the action he wished to take. It is more difficult to extend this interpretation of the dream into a larger context ("to magnify it" using the model of the news photo) and show that this dream means something generally true, say, about language and insight psychotherapy. True, a skillful storyteller could use the dream and its metaphor of the dictionary to expound numerous story lines. For example, we might suggest that the dreamer is a "person of the book" and speculate that it represents the Bible, a reference to Jewish and Christian patriarchy.

At some point, we go from plausible interpretation to wild speculation. Where is that point? To use clinical terms, when does a thread of interpretation deviate from reasonable and well reasoned and become extreme and paranoid? We cannot use all possible linkages between one metaphor and another (such as the book), because the number of routes is infinite. Without some constraint, interpretation becomes a game, like "Doublets," a word puzzle invented by Lewis Carroll. In this game one transforms a word into its opposite, such as DEAD into LIVE or BLACK into WHITE by changing only one letter at a step. If we take enough time and enough steps we can always find a way to transform one thing into its opposite. Fulfilling the alchemist's dream, we can transform LEAD into GOLD (i.e., lead > head > held > geld > gold).

A similar inversion occurs in *Hamlet*. The young prince taunts Claudius, who has ordered Hamlet to leave Denmark and flee to England because he has murdered Polonius. Hamlet turns and addresses King Claudius:

> *Hamlet*:   Farewell, dear mother.
> *Claudius*:   Thy loving father, Hamlet.
> *Hamlet*:   My mother: father and mother is man and wife; man and wife is one
>   flesh; and so, my mother. Come, for England!
>
> —*Hamlet*, act IV, scene iii, lines 56–59

Like Hamlet in this pointed attack on his mother's hasty marriage, some interpreters are tendentious. They seek at all costs to support a particular political or religious point. It takes only a few steps to turn BLACK into WHITE and for Hamlet to turn FATHER into MOTHER. We know that there are errors of the extremes: those who deny all possible meanings to the dictionary-clobbering dream do so out of obedience to their faith, not by reasoning. At the other extreme, those who deny any limit to interpretation run into trouble sooner or later. Our question, again, is when does that happen? Where is the tipping point?

When Freud took the interpretative path in *The Interpretation of Dreams*, he named one boundary of interpretation "the dream navel": "There is at least one spot in every dream at which it is unplumbable—a navel, as it were, that is its point of contact with the unknown."[55] This is a fascinating metaphor. It evokes a sense that the dream begins and is given birth within the self, in the deep recesses of the body or the mind. Like the newborn baby, dreams emerge still tied to their maternal origins. Using associated metaphors we can tie the unconscious part of the mind to the maternal sea, to the deep, and to the primal waters. One could continue along this route and argue that dreaming and the dream are therefore maternal and feminine. Or, one could take another, equally plausible route, and argue that the navel represents separation from the mother.

The navel is precisely not tied to the original source. Rather, it is a sign of difference and otherness. In classical Freudian theory, the father stands for reality testing, for the force that drives the child away from the maternal hold and toward the external world of difference. The father intrudes upon the mother-infant dyad and severs the primal unity. As messages linking one part of the mind to another and to the external world, dreams link inner and outer and define the two realms. Linkage and the bridge between unconscious and conscious is the province of the phallus. The punctuation mark the colon (:) is so named because it refers to the "member of phallus" and it links two parts of the sentence together, or so they say.[56] Thus dreaming and dreams are paternal and masculine.

The plasticity of metaphorical thinking is its charm. Yet, it is also a fundamental limitation. We cannot rely upon it to generate ideas or hypotheses that we can test against any set of facts. The "dreams-are-maternal" group can always find new metaphorical support for its camp, just as the "dreams-are-paternal" group can counter with its own metaphorical linkages. A third group might affirm *both* claims, but at a more synthetic level, and argue that male and female are themselves artifacts — constructed meanings — that function for our culture the way that dreams do for the individual.

On the one hand, we feel that metaphors are essential parts of self-understanding. They are not nonsense; they are not trivial; they are not to be discarded. It seems true that metaphors and metaphorical reasoning are essential to poetry, religion, and psychotherapy. On the other hand, neither are metaphors ordinary bits of knowledge. To use the model with which I began, in the natural sciences we seek ways to go to successive levels of complexity; in the interpretive disciplines we go horizontally, across skeins of potential meanings.

LEARNING ABOUT THE SELF: NEW HORIZONS

Thinkers like Ludwig Wittgenstein and Sigmund Freud succumb to the urge to find the essence of things by looking at them with special intensity. In this chapter I suggest that this urge stems from a universal fantasy that we can examine our minds and discover truths hidden there. This is a fantasy because it is both plausible and driven by wishes. It is plausible because we have access to our mind in ways that others do not. We can lie, for example. We learn at an early age that mother is not omniscient: she believes my lie when I said that I brushed my teeth. "Deep introspection" is a fantasy because it is driven by our wishes to believe that our inner world is as extensive as the outer world, that the mystery of consciousness, for example, and our inability to imagine its demise prove that a part of us immortal. To use a metaphor from the previous chapter, this fantasy includes also the belief that we can look within the self, that we can go deeper into the self just as scientists go deeper into nature by examining finer and finer details.

For all its rich evocations, Freud's commitment to this model of intellectual development made reality testing and ordinary person-environment interactions puzzling. I conclude this chapter by reviewing briefly two additional sources of reflection upon perception. The first resource is the stunning success of recent efforts to create machine-aided forms of perception for hearing- and vision-impaired persons. Using remarkably little information derived from electronic sensors, people can learn to see using data relayed to sensors on the tongue, for example, or on the skin of the back. As Paul Bach-y-Rita notes, while the brain can discern visual patterns using minimal data (as little as 114 bits/second in some cases), patients using computer-aided devices must act and react to their environments. That is, they must employ feedback loops present in the brain's vast capacity to realign itself using new data inputs.

The second resource derives from F. A. Hayek's lifelong criticism of central planning and his philosophic defense of the free market. Putting aside the ways in which one might idealize market mechanisms, Hayek's core insight was that free markets work better than any alternative because they increase the total amount of information available to both producers and consumers. Complex markets, like stock exchanges, are seismic indicators of what millions of people are thinking and doing in real time about matters of profound importance. In simplified terms, markets and other democratic institutions work better than central planning because they offer feedback data on the effects of unforeseen events. To put this another way, markets let us learn the consequences of decisions.

SEEING WITH THE BRAIN

A recent article in *Trends in Cognitive Sciences*, "Sensory Substitution and the Human-Machine Interface," carries this abstract:

Recent advances in the instrumentation technology of sensory substitution have presented new opportunities to develop systems for compensation of sensory loss. In sensory substitution (e.g. of sight or vestibular function), information from an artificial receptor is coupled to the brain via a human-machine interface. The brain is able to use this information in place of that usually transmitted from an intact sense organ. Both auditory and tactile systems show promise for practical sensory substitution interface sites. This research provides experimental tools for examining brain plasticity and has implications for perceptual and cognition studies more generally.[57]

Its authors, Paul Bach-y-Rita and Stephen W. Kercel, summarize more than thirty years of research devoted to developing electronic devices to substitute for sense-gathering systems, like that of normal eyes or normal ears. People whose eyes or ears are damaged can no longer receive sensory data through the usual biological channels. Without the aid of supplementary devices they cannot see or hear. But this does not mean that they have lost the ability to discern visual or auditory data. As Bach-y-Rita and Kercel say, "Persons who become blind do not lose the capacity to see. Usually, they lose the peripheral sensory system (the retina), but retain central visual mechanisms." This means that if one can supply a sight-deprived person, for example, with sensory data generated by a digital camera he or she can learn quickly to process that data stream and see.

This capacity illustrates the brain's ability to remap itself. Clearly the incoming data (from the tongue, for example) begins from an unusual place, yet they reach the brain structures typically associated with processing visual data.[58] Equally stunning are the discoveries that the brain can process patterns of visual data when they are displayed on the tongue, or on the back, or on other tactile surfaces. How this happens is not clear and is the subject of intense research. With these devices (which transmit very little data) even persons sightless from very early ages can acquire vision.[59]

Philosophers might challenge the claim that such machine-human interactions constitute real seeing. In response, Bach-y-Rita and Kercel note that if a machine lets a formerly sightless person receive data on the tongue, for example, that is structurally and systematically related to optically generated data from a camera and if the person can strike a falling ball that is vision. Like J. J. Gibson, Bach-y-Rita and Kercel, drawing on M. J. Morgan, cite the example of the horseshoe crab, which sees using unusual light-data gathering devices (or "eyes"). That some of the crab's photoreceptors are under its shell does not mean that it does not see.[60]

Using other devices, researchers have used gloves containing contact sensors and routed the data to the forehead of a patient whose leprosy had deprived him of normal touch: "After becoming accustomed to the device, the patient experienced the data generated in the glove as if they were originating in the fingertips, ignoring the sensations in the forehead."[61]

After training, subjects using these devices do not experience the sensation arising at the point of contact (on the tongue or back, for example). Rather, the devices merely transmit the visual or the auditory or touch data. (Bach-y-Rita and Kercel note that blind persons have used Braille to "see" books for more than a century.) Other researchers have found ways to substitute auditory signals for visual data. Using an audio device, P. B. L. Meijer trained subjects to see by linking the height of objects with pitch, and brightness with loudness when the subject scanned from left to right: "This yields a typical resolution of 60 by 60 pixels at a rate of about one frame-per-second."[62]

Early attempts to create human-machine-interfaces led to the problem of overloading patients' sensory systems. Trying to achieve maximum fidelity, researchers used devices that generated too much data. They learned that their first task was to reduce the amount of data produced and to make that data precise: patients are not "seriously impaired even in situations where a large proportion of the input is filtered out or ignored, but they are invariably handicapped when the input is very impoverished or artificially encoded."[63]

J. J. Gibson, a distinguished American psychologist, emphasized the centrality of the animal's response to its environment—that by moving through a given set of visual affordances, for example, the animal extracts from the systematic change in the visual array crucial information about the change in the visual angle, for example. An animal with even poor vision (or a human being relying upon a human-machine-interface) can negotiate its environment if it is permitted to act and then react to changes in the visual array. As Gibson never tires of saying: we cannot walk around hallucinations (or dream images, for that matter).[64] By walking around or moving in some other way, the animal employs these vastly complex feedback mechanisms.[65] Precisely which parts of the brain and which brain systems are involved in these marvelous events are not yet known. Brain imaging using PET scans and other forms of enhanced imaging devices have yielded some clues, but nothing is yet definitive.[66]

LEARNING FROM THE MARKET:
REASON AS AN INTERPERSONAL PROCESS

Friedrich August von Hayek (1899–1992) is best known for his work in the so-called Austrian school of economics led by Ludwig von Mises. Hayek extended Mises's critique of central planning and carried on a lifelong debate with Alfred Keynes, the premier economist of his times. While he wrote on the business cycle and its fluctuations, Hayek is better known for his works on the rise of socialism and his essays on political theory, philosophy, and legal theory.[67] It may seem odd to leave the folds of psychology and talk about an economist. Hayek did not restrict himself to technical matters of economic theory. He also focused upon broader questions of political theory and the status of human beings as autonomous persons. Hayek's heroes are De Tocqueville, David

Hume, Adam Smith, Edmund Burke, Lord Acton, and others who defend the idea of individualism as opposed to the celebration and assumption of a pure rationalism, a system from which one can deduce the rules of society.

Hayek opposed René Descartes (1596–1650) and Jean-Jacques Rousseau (1712–1778) because each championed the ideals of an individualism that grows, Hayek says, inevitably, into collectivism. He cites a passage from Descartes's *Discourse on Method*, Part II (1637), where Descartes claims that perfection typically derives from the handiwork and mind of a single creator. By perfection, Descartes means using deductive methods, modeled in mathematical reasoning to arrive at certain truth:

> These long chains of reasoning, so simple and easy, which geometers customarily use to make their most difficult demonstrations, caused me to imagine that everything which could be known by human beings could be deduced one from the other in the same way, and that, provided only that one refrained from accepting anything as true which was not, and always preserving the order by which one deduced one from another, there could not be any truth so abstruse that one could not finally attain it, nor so hidden that it could not be discovered.[68]

When kings and philosophers attempt to apply this model of reasoning to nonmathematical tasks, problems emerge. If we accept Descartes's claim that this mode of reasoning lets us deduce all possible truths, we have no need for raw experience, much less the voice of plebeians and others who cannot comprehend the majesty of mathematical rigor. When this kind of Cartesian rationalism is applied to governance, totalitarianism must result because rationalism permits no room for the muddy stuff of evidence and market driven responses.

In opposition to Cartesian rationalism Hayek insists upon economic facts:

(1) The extreme complexity of any large-scale economic and political structure. Because these are the result of millions of persons making thousands of decisions no single mind, much less verbalized model or simplistic theory of history can capture these interactions accurately.

(2) Human beings are inherently limited, self-centered, and non-rational. Their moral lives especially are not the products of reasoning alone.

(3) Human beings, therefore, will not comport or bend themselves to "reasoning" as outlined in any fully articulated treatise.

The rationalist tradition, exemplified by Descartes and socialist and collectivist theory, tends to disdain social institutions that have merely evolved.[69] Drawing upon a long history of social critique and rhetoric, rationalist traditions champion a certain kind of Reason with a capital R that is "always fully and equally available to all humans and that everything which man achieves is the direct result of, and therefore subject to, the control of individual reason."[70] Hayek disagrees. He argues that such reason does "not exist in the singular, as given or available to any particular person but must be conceived of an interpersonal process iñ which anyone's contribution is tested and corrected by others."[71]

Hayek notes that persons must be governed by general principles that permit actors to know the basic rules of competition yet avoid totalistic claims to control the outcome of competition. He warns against perfectionism and the overvaluation of reason, that is, conscious, technical, means-end rationality: "Our submission to general principles is necessary because we cannot be guided in our practical action by full knowledge and evaluation of all the consequences. So long as men are not omniscient, the only way in which freedom can be given to the individual is by such general rules to delimit the sphere in which the decision is his."[72]

With competition, price movements, and price communication comes increasing information about factors significant to each actor (customer and seller). Hence, in any such encounter, maximizing freedom of information maximizes efficiency. The amount of information increases for everyone. This means that everyone makes better decisions, increases his or her productivity, and therefore increases private and public wealth.

Some see Hayek as too conservative politically, and many conservative politicians and savants have sought refuge under this name. However, Hayek criticizes most harshly those who favor central planning of any type—whether by the state or by giant corporations. For both institutions depend upon two convictions: that high-ranking people know better than others and that their superior intelligence gives them deeper insights into markets of all types.

I have argued against both of these convictions. I cite Hayek as yet another way to make the same point: to the degree that the mind is cultural, it is organized horizontally, by associative networks. Free markets, free exchange of ideas, the open university, scientific method, scholarly honesty, and a free press—to name only a few characteristics of the political system we cherish—are structured to increase the flow of information and ideas across vertical boundaries. We return to Swammerdam and the frogs he dissected in the seventeenth century. We note that everything he did was public and replicable. This fact is crucial for the development of scientific research. It means that anyone— Asian, female, Black, or atheist, to name groups who had no access to scientific training in his time—can, given proper instruments, judge Swammerdam's findings by repeating his experiment. As Freud reminded the students in his introductory lectures in 1915–1916, they too had seen the shortening of a muscle as a result of the stimulation of its nerves.

In their best moments, scientists and scholars reject claims of natural hierarchy. The institution of blind review of manuscripts, for example, recognizes that our too-human bias for or against famous colleagues impairs our judgment. By submitting themselves to rules of evidence and blind review, scientists and scholars expand the notion of reason and mind. The latter two name, in this sense, a superconscious order, a collective ability to restrain mere narcissism in favor of discovery.

7. HIGH ART AND THE POWER TO GUESS THE UNSEEN FROM THE SEEN

DOES HIGH ART CONVEY KNOWLEDGE?

"There's my hometown!" I shout. I catch a murder mystery set in Portland, Oregon. It's about a pretty woman, a handsome man, and a villain hell-bent on harming her. Also, her outdoor furniture, it happens, is the kind I'd bought a month before. "Look!" I announce to the room, "There's Sellwood Bridge! I rode across it every day going to college." These insights are pleasurable to me alone. Others, no matter how kindly disposed, do not find scenes of a bridge and rainy streets interesting. My fascination with Portland defines sentimentality. Like other people's photos of their grandkids, my recollections are boring. A similar problem plagues astrology and spirituality; we sympathize with the urge to discover how the universe pertains to us, but these urges cannot, by themselves, rise to the level of art. What do they lack? I will suggest that arts convey knowledge through metaphorical unions, by finding novel patterns, not through the pursuit of hidden meanings. To do this I consider two Victorian authors, John Ruskin and Henry James. I conclude by returning to the issue of magnification and the fantasy of infinite depth evident in three movies and the misguided theories of a distinguished physicist.

Reflecting upon sentimentality, John Ruskin, a noted Victorian critic, distinguished between three levels of art: fancy, emotionally driven (pathetic), and the arts of high imagination. Fancy seeks to entertain and offers no pretense of believability. Pathetic art illuminates how people act under the sway of intense emotions and so becomes believable: "By manipulating a portion of reality which both speaker and listener share, the pathetic fallacy allows one to glimpse the passions within the consciousness of another human being."[1] Even the best of pathetic arts, such as lyric poetry, are limited because they are restricted to the poet's experience. George Landow writes: "When the poetic speaker feels happy, everything appears perfect, all rings with joy; when he experiences grief, everything appears colored by his grief—the waves either reflect it or cruelly mock it."[2]

Ruskin does not elevate pathetic literature, which conveys a person's strong feelings, to the highest level because it does not locate that suffering in a larger pattern of life. It pertains only to those who share the poet's idiosyncratic history. High art must transcend individual experience and present a more complete pattern of life as a whole, in which persons and individual emotions are distinguished from one another. For these reasons, Ruskin admired Dante:

> Dante's broad view of life will not permit his deep sympathies to distort reality; and when he "describes the spirits falling from the bank of Acheron 'as dead leaves flutter from a bough,' he gives the most perfect image possible of their utter lightness, feebleness, passiveness, and scattering agony of despair, without, however, for an instant losing his own clear perception that these are souls, and those are leaves."[3]

Ruskin champions high art and argues that it conveys new knowledge. The most accomplished artists, like Homer, Dante, and Shakespeare, recognize the organizing power of emotionality but do not immerse themselves in it. Great artists have vision that is clear, of high imagination and calm consideration. They do not present transitory emotions; they portray turmoil within a larger framework of clarity and poise. The greatest poets surmount the flux of consciousness, as Landow puts it. In so doing they provide equanimity and knowledge that makes art the equal of the sciences. In these moments of communication about our shared destiny, high art transcends fancy.

A contemporary example of high art is Meredith Willson's funny, intelligent description of life in 1912 Iowa in *The Music Man* (1957). With economy and wit, Willson portrays prewar Iowa (River City) much as James Joyce portrayed Dublin on June 16, 1904, in *Ulysses*. The charm of *The Music Man* resides in its music; its force resides in Willson's critique of his sentimental longing for 1912 Iowa, when cigarettes were illegal.

In the play's opening song, "Rock Island," the salesman barks out, "Gone, gone, gone." We recognize lamentation:

Gone, gone, gone with the hogshead, cask, and demijohn;
Gone with the sugar barrel, pickle barrel, milk can;
Gone with the tub, and the pail, and the tierce.

Three hammer blows of "gone" are followed by the exact names of three sets of three containers. Each name designates an experience that is lost forever. Like all genuine mourning, this verse names the precise thing, in its precise location, and declares that it is gone. By using archaic words—*demijohn* and *tierce*—Willson tells us that his boyhood home, the world in which these funny words made sense, is vanished.[4] This courage to name the thing that is lost is a feature of tragedy and, in this case, a Broadway musical.

TRAGEDY AND MOURNING AS PROGRESS

Affirming the value of tragedy are psychoanalysts who champion persons willing to recognize the permanency of loss. Freud made this explicit in his masterful essay "Mourning and Melancholia."[5] The work of mourning, or grief, requires us to acknowledge that the beloved is gone. Freud says that we must then give up our wishes, piece by piece, that the beloved Other is not gone. Because we retain these wishes, as it were, in separate parts of our minds, we must examine them one by one. We cannot obey reality all at once:

Each single one of the memories and expectations in which the libido is bound to the object is brought up and hypercathected, and detachment of the libido is accomplished in respect of it. Why this compromise by which the command of reality is carried out piecemeal should be so extraordinarily painful is not at all easy to explain in terms of economics.[6]

This painful work marks the beginning of new capacity, what Freud and other moralists call the adult attitude. The latter is an attitude of tragic self-acceptance of the limitations of our childhood wishes. Without this kind of work and suffering, the person cannot achieve a new kind of freedom.

While we acknowledge the brilliance of Freud's essay, must we also share its metapsychology, that is, the notion that mourning is internal *work* and that it has an energy component? Is mourning an internal process, best captured in Freud's metaphor of "hypercathecting and detaching the libido" from internal images? We know the source of this metaphor: it derives from Freud's commitment to the biological models announced in "A Project for a Scientific Psychology" and revived in *The Interpretation of Dreams*. With his customary dazzle, Freud ties these diverse discourses together. When Freud says, "Why this compromise by which the command of reality is carried out piecemeal should be so extraordinarily painful is not at all easy to explain in terms of economics,"[7] he means that his commitment to the notion of psychic energy requires him to say that psychic pain consists of an excess of energy. This is driven by his hope to link clinical observations to nineteenth-century notions of energy.

Is this language of energies and structures necessary? Consider a comment by Melanie Klein on mourning. She describes a woman whose child has died and who recalls, unconsciously, memories of an "early dread of being robbed by a 'bad' retaliating mother is reactivated and confirmed. Her own early aggressive fantasies of robbing her mother of babies gave rise to fears and feelings of being punished, which strengthened ambivalence and led to hatred and distrust of others. The reinforcing of feelings of persecution in the state of mourning is all the more painful because, as a result of an increase in ambivalence and distrust, friendly relations with people, which might at that time be so helpful, become impeded."[8]

Klein shows how dread from one's childhood is added to one's current misery and loss. She deepens our understanding of pathological mourning by noting the likely connections between her patient's childhood beliefs about her bad mother and adding the key insight that that catastrophe turned, in part, on the young girl's projective mechanisms. While all children must deal with similar ideas about mothers, bodies, and envy, not all children acquire such tenacious pathological beliefs. We suspect that no adult intervened to challenge and undo the child's misconceptions. That failure in interpersonal education is repeated when the adult woman cannot bring herself to confide in her friends during this period of suffering.

Faithful to the Freudian metaphor (or model) of the inner world and its structure, Klein repeats the assertion that the painfulness of mourning serves "not only to renew the links to the external world and thus continuously to reexperience the loss, but at the same time and by means of this to rebuild with anguish the inner world, which is felt to be in danger of deteriorating and collapsing."[9] Again, we accept this helpful metaphor because it helps us understand the patient: that mourning feels like work, that the views of oneself are as if one were naked, or collapsing, or destroyed.

But Klein means more than this. She claims that there really are internal structures and that the goal of therapy is to help the patient rebuild them or to help her build better ones, as Freud said in "Mourning and Melancholia."[10] We find much value in Klein's clinical description. It illuminates a kind of suffering that we recognize in our patients and in ourselves. That is, because the woman who has lost her child also believes that she had harmed her mother, the new loss reawakens those archaic beliefs. Does it add to this beautiful description to say that the woman is trying to "rebuild with anguish the inner world"?

As we have seen, Freud began with neurological theories that portrayed the mind as a complex, biological machine. Machines require organization and a supply of energy, just as bodies need a constant source of energy. It follows that the mental machine requires some form of energy that permeates the nervous system. Health is its proper functioning; disease, especially when there is excessive action or excessive pain, is the opposite: "We may say that hysteria is an anomaly of the nervous system which is based on a different distribution of excitations, probably accompanied by a surplus of stimuli in the organ of the mind."[11]

Unlike machines, the mind operates with entities called thoughts, not raw energies. To reflect this fact Freud proposed that pools of energy (or excitations) are "distributed by means of conscious and unconscious ideas."[12] The problem of consciousness presents special problems. Freud confronted this issue in his preface to a book by Hippolyte Bernheim: "We possess no criterion which enables us to distinguish exactly between a psychical process and a physiological one . . . ; for 'consciousness' whatever that may be, is not attached to every activity of the cerebral cortex, nor is it always attached in equal degree to any particular one of its activities; it is not a thing which is bound up with any locality in the nervous system."[13]

The mystery deepened. For, as Freud noted: "In its paralyses and other manifestations hysteria behaves as though anatomy did not exist or as though it had no knowledge of it."[14] Therefore, if organic lesions cause hysteria, these lesions are shaped by the patient's mental representations of the patient's body. To retain his physiological commitments, Freud claims that these lesions must be the result of the upsurge in a surplus of excitations. And, therefore, psychotherapy must be the "abreaction of these surpluses."[15]

In the Freudian view, the repetitive behaviors that typify mourning signify an inner process of detaching libidinal ties: "The mourner's problem, in this conceptualization, is essentially an economic one. His libido is bound up in a departed object. He experiences no satisfaction and may feel quite empty. Only when his finite quantity of libidinal energy is detached from the lost object and can be invested in a new object does he recover." This brief, elegant, yet detailed statement has become one of the most familiar and influential in Freud's writings. It has been the source and foundation for all later psychoanalytic theorizing about mourning.[16]

A century of reflection upon the economic point of view has let us distinguish the clinical insight about mourning, at least in typical Western contexts, from the concept of *psychic energies*. While there is much to gain by observing how patients react to the loss of loved ones, there seems little to gain by adding to these descriptions comments about the flow or retardation of energies that cannot be measured. On the contrary, a commonplace of many sophisticated comments on mourning by much-esteemed analysts is to show how mourning is advanced by ancient traditions of religion and poetry. Neither of these depends upon a theory of mental energies or upon a notion that mourning is the decathexis of one's ties to the beloved. Salman Akhtar, a psychoanalyst and a poet, describes the "cultural ointment of poetry": "Poetry also seems to have an uncanny and direct access to the deepest layers of psyche, which is where mental pain usually originates and resides. If reading prose is like taking pills for arthritic ache, then reading poetry is akin to that of a steroid injection directly into an inflamed joint."[17] Reading and writing poetry, he notes, can tell us how we feel. Akhtar cites renown poet Seamus Heaney, who noted that typically the poet writes, then discovers what he or she has written.[18]

While I agree with these descriptions of poetry and the goals of psychoanalysis, what is added to these accounts by citing Freud's notion of the economics, which is the energy aspects of psychic pain?[19] Neither writing poetry nor telling one's story is a microscopic process. In a sensitive description of Anna Freud's response to the death of her father, Robert Gaines remarks upon her actions. First, though, he affirms Freud's economic thesis: "The mourner's problem, in this conceptualization, is essentially an economic one."[20] Is this a useful way to talk?

Mourning is a long, painful process of remembering feelings and images of ourselves and the beloved. In this sense, mourning is the inverse of falling in love. To fall in love with another person is to construct multiple versions, that is, multiple images of us together in future times (and to find archaic wishes, dreams, and fantasies reawakened). Forsaking our investment in those images of future happiness is painful because we have to undo the imagined worlds we have constructed. Like a novelist forced to destroy her newest, unpublished work, mourning asks us to dissolve a world we have created and populated with our deepest feelings.

Psychoanalysts long ago described a parallel spectrum to fancy, emotionality, and high art. This parallel spectrum stretches from the daydream, with its comic-book stories of adventure and wish fulfillment (where everything is possible), to the pathos of the soap opera (where the story is merely improbable), to the insight and self-awareness of tragedy (where life itself reappears highlighted and deepened.) A fine instance occurs in a contemporary comic-tragic masterpiece, *The Sopranos*. An American mob boss laments his fate, asking what kind of person he must be if his own mother conspired to kill him.[21]

Freud's mixed language, as Paul Ricoeur put it, has direct consequences for clinical theory and technique. By emphasizing the concept of *internal work*, Freud champions a view of the patient's ego as examining its internal world, the world of virtual images associated with charges of energy. This gives primacy to the patient's accounts of his or her thoughts, especially dreams and other spontaneous products, because theory tells us that they emerge from internal struggles between competing parts of the psyche. Consistent with this view, Freud required the analyst to deduce the nature of these internal conflicts and to help the patient gain insight—that is, to see *within* the self. This takes us back to the metaphor of depth. This metaphor feels accurate because surely *my* memories reside *inside* of me. But this is a misleading deduction. While my memories and feelings occur within me, they are shaped by and amplified by art (in all its forms). Music, song, poetry, and other public arts structure my inner world the way that language organizes my mind. Ritual, music, and other public arts make mourning possible. Added to these ancient resources is the modern art of psychotherapy. Therapy is a deeply ritualized form of interaction that, at its best, offers the possibility of new forms of mourning and reconciliation of our wishes with our fate.

THE POWER TO GUESS THE UNSEEN FROM THE SEEN

We share a common ability to deduce the future from hints available in the present. In order to survive in the novel environments presented to us, we must find pattern in the midst of change. In more exact language, vision and the other sensory systems must supply us with sufficient sense of sameness that we can navigate in arenas we've never seen before. With regard to social environments, infants and children must learn the deep structures of language and the analogous structures of social interaction. The latter means that a culturally sophisticated actor can understand and predict patterns of novel social interactions. If knowing a language means knowing how to make entirely new sentences—or understand novel utterances by others—then knowing how a particular form of life works means we can grasp novel versions of it.[22]

Henry James said something similar in "The Art of Fiction." He described how an English novelist, "a woman of genius," received much praise for her novel about a French Protestant woman. How had the English woman learned so much about this foreign topic? "These opportunities consisted in her having once, in Paris, as she ascended a staircase, passed an open door where, in the household of a *pasteur,* some of the young Protestants were seated at table round a finished meal. The glimpse made a picture; it lasted only a moment, but that moment was experience. She had got her impression, and she evolved her type." She knew what youth was and what Protestantism was; she also had the advantage of having seen what it was to be French; so that she converted these ideas into a concrete image and produced a reality. "Above all, however, she was blessed with the faculty which when you give it an inch takes an ell, and which for the artist is a much greater source of strength than any accident of residence or of place in the social scale. *The power to guess the unseen from the seen, to trace the implication of things, to judge the whole piece by the pattern,* . . . this cluster of gifts may almost be said to constitute experience."[23]

While James ascribes this cluster of gifts to men and women of genius, it occurs in all persons, certainly in infants and children, who must learn to "guess the unseen from the seen." That is precisely the task of learning to live in a world of constantly changing appearances. This capacity cannot be mere fantasy, for fantasy is disconnected from observation. The novelist's ability to grasp the nature of French Protestant youth could not have derived from her daydreams. It derived, rather, from her observations of the young woman in the house and her ability to extrapolate from the other's demeanor, and a hundred other clues, the nature of the girl's lived world, that is, the worlds in which the young woman lived.

James tells us that the novelist knew what it was to be a young woman, to be Protestant, and to be French. From this information she created a "concrete image" of a young, French, Protestant girl. Readers who knew such women validated the novelist's intuitions by inquiring as to how many years she had studied them. James defends the novelist, and himself, of course, against such narrowmindedness. Earlier in the passage cited above, James notes that another young woman, "upon whom nothing is lost," might speak the truth about military men even though she had not served among them. This is not miraculous. On the contrary, slyly broaching the narcissism of military men, James says that even a young damsel of talent can comprehend their experiences and portray them in fiction. This moment of high art occurs when an observant person sees the seemingly superficial patterns that "men of action" display.

REALITY TESTING AS AN
INTRAPSYCHIC PROCESS

Freud affirmed what most scientists of his day believed: perception, especially vision, could be best understood on analogy with the numerous machines perfected in the nineteenth century. Vision must be the activity by which the animal "sees" the retinal image, just as the camera film registers the image passed through the camera's lens and projected back onto the plate. Vision is the processing of sense data, or the bare quantities that Freud described in "A Project for a Scientific Psychology."

There is no doubt that mourning occurs "inside" the mind in the sense that persons experience the emotional suffering tied to mourning as if it were within their bodies and the private theater of the mind. Our task is to figure out the best way to talk about this "inside." Leaning on the neurology of his day, and given his focus on the individual's self-reflection and spontaneous behaviors, Freud focused his genius upon the arena of self-reflection. In doing so he overemphasized the ubiquity of hallucination.

In an earlier study, I describe what I called the standard notion of perception in Freud.[24] Pinchas Noy gives a precise summary of this standard theory:

> The tremendous amount of information pouring into the apparatus is incessantly meaningless to the receiver until the data are processed and organized into meaningful schemata which can be identified by matching them up against corresponding schemata stored in memory. Almost any one of the various mental functions, such as perception, memory, orientation, or assimilation, can be carried out only after the raw data of input have been organized into some kind of meaningful and identifiable schemata.[25]

Noy expands upon this view of perception when he says that the "human computer" uses an internal program to manipulate raw data into representations. Noy is consistent; he refers to a phenomenological maxim, "Keine Gestalt ohne Gestalter." We may translate this German maxim as "no pattern without a patterner," which means that patterned experiences derive from the experiencing subject, not from the external world alone.

Anyone trained in Anglo-American philosophy in the last forty years will have read J. L. Austin's book *Sense and Sensibilia*, a critique of this notion of sense data. While Austin trains his criticism upon philosophers like A. J. Ayer and H. H. Price, who argue that perception is the perception of "sense data," his book pertains to Freud as well. Like these philosophers, Freud makes vision a *problem*: "The general doctrine, generally stated, goes like this: we never see or otherwise perceive (or 'sense'), or anyhow we never *directly* perceive or sense, material objects (or material things), but only sense-data (or own ideas, impressions, sensa, sense-perceptions, percepts, etc.)."[26] Austin exposes the metaphysical claims embedded in the notion of "sensibilia," especially the demand that we ignore differences among types of sense experience and reduce everything to one of two types: either veridical perception or hallucination (or delusions)— Austin calls this a bogus dichotomy.

During a dream we believe that what we're experiencing is real, that the giant Gila monster will crush our car. During a dream—when we're immobile—we find it difficult to distinguish dream-states from ordinary states. Freud argues throughout "A Project for a Scientific Psychology" and the *Interpretation of Dreams* that these dream-states occur because our ego mechanisms have regressed to the earliest states of wish-driven perception, where hallucination dominated. Just as the hungry infant imagined the breast where it wished it to be, so too during dreams we (sometimes) hallucinate the presence of wished-for objects.

Austin asks: Does the dreamer see illusions or does the dreamer hallucinate? He answers that the dream-state is one of *dreaming*, that is, a normal experience that people can quickly learn to distinguish from other states of mind. Freud, Ayer, and other proponents of the notion of sense data hold that human beings cannot distinguish "qualitatively" between hallucinated images and "real" images, that is between merely imagined objects and the vision, for example, of ordinary objects we can see under ordinary circumstances. Here, J. J. Gibson, a distinguished American psychologist, comes to our aid.[27] As Gibson notes, hallucinations, dreams, and conjurors' tricks depend upon us *not* moving our head during the process. When magicians make it seem that they're sawing a lady in two, we are seated, facing them directly. We cannot get up and wander behind the stage or even to the edge of the stage because then we'd see how the trick is done.

Addressing himself to the same question, Freud committed himself to locating the activity of reality testing within the psyche. Believing that hallucination was the original status of the infant, Freud and generations of analysts have labored to explain how reality testing occurs within the psyche. Reality testing must be an achievement because theory tells us that the ego spontaneously seeks regression and avoidance of the external world. By making hallucinations the original tendency of the mind, Freud was able to link myriad data about other forms of regression, such as that seen in dreams, psychotherapy, war-induced trauma, and archaic religion. Each of these provided yet more evidence that human beings and human culture must resist the pull of regression.

Because he held that reality testing occurred inside the mind, Freud had to explain how the ego distinguished between imagined ideas and perceptions of reality. His answer was that "real things" are distinct because we attach to them "qualities." This distinction appears in the third part of "A Project for a Scientific Psychology," where Freud addresses two questions: how are qualities remembered, and why are speech associations (hearing) primary? To account for normal psychological processes (as opposed to pathological ones) he offers a theory of the neurone (neuron). Freud's explanation is laconic and difficult to condense further.

Unlike idealist philosophers and psychologists, Freud says the mind became complex because it evolved to protect the human animal. Freud's famous aspersions upon the burdens of civilization are literary formulations of this thesis. To control itself and to use thought rather than action, the ego must increase the amount of bound cathexes (that is, bounded energy) available to it. (Hence this problem dominates the rest of Freud's account.)[28]

Thing presentations (or ideas, German *Vorstellungen*) are those parts of the perceived image that are generic. The child perceives its mother as a "thing" (object) that manifests characteristics typical of mature females. To perceive its mother the child must recognize her distinctive features, her distinctive qualities. This is the role of word-presentations. Word-presentations are linked biologically to speech, and speech is a form of motor enervation.

Primary process thought fails to carry out this form of judgment. For primary process thought treats similar objects as if they were identical, and therefore the ego initiates discharge (action). The neonate nurses with any vaguely appropriate object. The dream considers babies, penises, and feces as interchangeable entities. An analytic patient deep in a transference neurosis responds to the analyst as if the analyst were mother. Freud does not say that the infant judges the perception of a breast against the word-image "breast." The infant does not use language to distinguish safe objects from dangerous ones. Rather, the child's capacity for language marks it as having achieved a level of secondary-process functioning. This capacity distinguishes acting upon the perception of objects that are similar to one's wishes and acting with rational judgment.

I have focused upon Freud's theory because it exemplifies a commonsense belief: that the mind is an internal organ of some type and that by looking inward we can better understand its functioning. Closely linked to this belief is the perplexing nature of consciousness: we alone can "know" our feelings, yet this kind of knowing is not identical to knowing about the so-called external world. Dominating Freud's work—and that of numerous psychologists who followed him—is the task of explaining how these two forms of knowledge differ.

In this book I do not solve that problem, but I do suggest that the *objects* of knowledge differ. We saw this in the slide show "Magnification." The apparently trivial point made there again and again is that natural objects can be studied and magnified in numerous ways. Even at a magnification of 70,000x we can discern structures in the images of the viruses. Where we can discern structure we are discerning information, not randomness. And, where we can discern information we can gain knowledge.

This is not true of cultural objects, like the images of Abraham Lincoln and Susan B. Anthony. Most people have no notion of who or what the slides portray until we reduce magnification to 4x or 3x. Even this is too generous, for as any art critic or music critic will attest, the conditions for a proper viewing or proper listening are not plus or minus 2x. Seeing Michelangelo's statue "David" at one-half size is not the same experience as seeing the original in its full size; listening to *The Magic Flute* at double the usual tempo ruins the opera. This means that we cannot apply the rules of natural science to cultural objects; we cannot magnify them and we cannot learn new things about them by submitting them to this kind of vertical analysis.

To the degree that self-understanding—and consciousness—is a cultural object, it will not yield to vertical analysis, that is magnification, either. To pursue magnification in all its guises, including deep introspection, is to believe that we can discover new truths about the self by looking inward, toward deeper and deeper levels. This, I think, is impossible. More than that, by seeking to examine and magnify data about the self, we quickly generate noise, not signal. When I show the "Magnification" slides to large groups, I ask everyone to shout out what they believe each object is as soon as possible. This introduces a lively element of competition and an inadvertent projective test. For until we get to 4x or 3x, slides of the cultural artifacts are mainly random visual noise, like that found on television sets where there is no signal. The absence of a genuine, meaningful signal does not prevent creative people from seeing things.

For example, slide E happens to have no content: I merely asked my research assistant to make a blank. He did so, but happened to include in his preparation a less than perfectly blank background. Thus when I show E to large groups, someone always "finds" in it a strand or blotch in which they discern a face or a sign, etc. This is harmless fun, unless something valuable is at stake. If, under similar circumstances of very low signal but very high anxiety—say in the United States in the early 1950s—we feel endangered by unseen, powerful forces, we will feel compelled to study minimal signs. When we lack genuine information we will project onto the noisy surface. Projection and paranoia are processes evident in the numerous movies. We consider three below.

LOOKING OUTWARD, THREE MOVIES

The idea that hidden, organizing powers operate beneath the perception of ordinary people brings us to yet another version of the paranoid system. For in the latter we find the same belief that behind seemingly normal events there is an evil consciousness, lurking, controlling, and plotting to harm the innocent. As children of Watergate and other notorious lies and cover-ups, we cannot exclude the possibility that sometimes there are those who plot against us. We see these issues emerge in three significant films from that period: *Blow Up* (1966) directed by Michelangelo Antonioni, *High Anxiety* (1977) directed by Mel Brooks, and *The Conversation* (1974) directed by Francis Ford Coppola. I cite these films because they take us to the heart of this book: How the unexamined urge to find hidden truths within cultural objects produces a noisy field upon which we project our worst fears.

BLOW UP

In *Blow Up* (1966),[29] a film inspired by Hitchcock's epic study of voyeurism, *Rear Window* (1954), and Michael Powell's 1960 violent movie, *Peeping Tom*, Michelangelo Antonioni made this point with panache. A famous picture on the soundtrack album shows the hero Thomas (David Hemmings) engaged in his swinging, hippie profession of fashion photographer and girl-chaser.

Sensational for its time, banned by Catholic censors, notorious for its sex scenes—including a sadistic romp between the hipster hero and two young women—*Blow Up* brought swinging London to middle-American movie houses. Those who saw the film in 1966 will recall many a late-night conversation about the movie, which some hailed as a masterpiece; others talked mainly about the naked women and the sensational murder mystery. The latter, captured in a beautiful, apparently empty London park, involves the photographer (whose name we never hear in the film) and his accidental encounter with a pretty young woman and her older boyfriend. Snapping a dozen photos of the couple, the photographer is surprised when the young woman (Vanessa Redgrave) follows him to his studio, demanding the negatives. This and her sexy style intrigue him; he placates her with a dummy roll of film, she leaves, and he develops the authentic film. In a series of brilliant scenes, Antonioni shows him make larger and larger blowups and finally discern what seems to be a man with a gun hidden in the shadows. Another photograph taken later, seems to show the body of the older man hidden in the park.

The screenplay was based on a short story, "Las Babas del Diablo" (The Devil's Drool) by Julio Cortázar, a noted Argentine writer heavily influenced by surrealism and Edgar Allan Poe. As one critic notes, "Having read quite a bit of Cortázar, I know that he enjoys wordplay, and I think he is suggesting with the title that the camera is a drooling devil—a lustful voyeur that is capable only of lifeless illusion and is ultimately impotent. In the story, Cortázar's narrator writes: *I think that I know how to look, if it's something I know, and also that every looking oozes with mendacity, because it's that which expels us furthest outside ourselves.*"[30] In the short story, set in Paris, an amateur photographer enlarges a negative of a woman whom he deduces is seducing a teenage boy to procure for a man who is waiting in a car. This introduces the theme of paranoid worry and doubt: "Nobody really knows who is telling it, if I am I or what actually occurred or what I'm seeing . . . or if, simply I'm telling a truth which is only my truth."[31]

Antonioni retains this element of doubt when he begins and ends the film with scenes of a "Band of Merrymakers"—a truck filled with mimes who accost everyone with their loathsome white faces and pseudohilarity. In the much-discussed final scenes, the photographer wakes to find that his film and all but one of the prints of the park scene have been stolen and the body in the park, which he saw in person when he went to investigate, is gone. (The one print that the thieves overlooked is too obscure to use as proof of a crime.)

As he said to his agent at a swinging, druggy bash the night before, he couldn't be sure if he'd seen the body unless he photographed it. The film ends with the photographer walking away from us in a grassy park, his camera swinging by his side. He passes the Merrymakers playing mime tennis. At first he and the camera ignore their game, but soon the camera (that is, Antonioni's camera) and we follow the flight of the imaginary ball. The ball is hit out of court, the photographer hesitates, then pursues it and throws it back into play; as he turns and walks away we hear the sound of a real tennis match. Then he disappears, and we are left with the metaphysical puzzle: what is real and what is not? Or, as Keats put it in Cortázar's favorite poem, "Ode to a Nightingale":[32]

> *Was it a vision, or a waking dream?*
> *Fled is that music:—Do I wake or sleep?*

Written around 1819, Keats's poem evokes a mood that most people can recall, certainly those who first discover the intensity of sexual longing and the poignancy of desire and imagination. Like other poems of his best work, Keats draws upon the taken-for-granted beauty of nature, especially the much-tended gardens and fields of England. His gorgeous language and meter reinforce the struggle between drowsy wishfulness, the vision, and wakefulness. The poem's final lines are questions about the poet's impulses; they are not answers, and they do not ask us to assent to extranatural claims. Keats's questions are rigorous; they sharpen our sense of difference between the two states of mind.

This contrasts to the paranoid attitude; in the latter we cannot enjoy the luxury of asking questions, much less enjoy Keats's odes. To be paranoid is to feel compelled to act. One is committed to a dangerous plan without knowing fully against whom or what one struggles. Within corrupt police states, like Argentina of the 1950s, perceptive people acted in ways that might appear paranoid—yet genuine evils lurked everywhere. In an interview in 1973, Cortázar described his coming of age politically through his novel *Rayuela* and its main character, Oliveira:

> I believe it is a profoundly optimistic book because Oliveira, despite his quarrelsome nature, as we Argentinians say, his fits of anger, his mental mediocrity, his incapacity to reach beyond certain limits, is a man who knocks himself against the wall, the wall

of love, of daily life, of philosophical systems, of politics. He hits his head against all that because he is essentially an optimist, because he believes that one day, not for him but for others, that wall will fall and on the other side will be the kibbutz of desire, the millennium, authentic man, the humanity he's dreamt of but which had not been a reality until that moment.[33]

Cuban Marxism and other solutions offered by the Left intrigued Cortázar because they helped him resist North America and its domination of Argentina. Magical realism and other forms of literature typical of South America are, in this sense, forms of protest; they refuse to accept that this world is all that can possibly exist. Antonioni shares a similar sense that the wild, capitalistic excess of fashion photography—and the hero's lust for money—destroy ordinary pleasures. Everything is measured by costs. (In the sex romp with the two young women, the photographer demands sex. The girls wish to use him to advance their modeling careers: it's a market exchange.)

Commenting about how he starts to write a film script, Antonioni describes a very different scene of two young girls that he happened upon when he was beginning a new movie:

> Two girls about nine years old are playing. One of them is going around the rotunda on a bicycle. The other one goes nimbly into a handstand, stiffening up in vertical position, her skirts falling over her face, her skinny legs straight up in the air. Then she lets herself fall over and starts again. They are poor girls. The one going around the rotunda on her bicycle yells, in sing-song manner,
>
> *"Oh such love, oh such suffering! . . ."*
>
> She disappears, then comes back.
>
> *"Oh such love, oh such suffering!"*
>
> It's early morning, there's no one on the beach except me and those two girls. No sound other than that of the sea and that frail voice crying love and suffering. For me, that was a film the rest of that day.[34]

Central to this memory is that the girls were young and poor; they entertained themselves by singing an ancient love song. Visually, the scene evokes a cinematic feeling because, as Antonioni tells us, each girl does something interesting. The one girl rises and falls in her handstands; the second girl rides her bicycle round and round. We hear her song, then she and the sound disappear, then both return when she comes round the rotunda. These scenes and their associations helped Antonioni realize his movie: "What I feel most is the memory of the moments which inspired me to conceive and write those things. Visits to certain places, conversations with people, time spent at the very spots where the story is said to take place, the gradual unfolding of the picture in its fundamental images, in its tone, in its pace: This is very important to me. Perhaps the most important time."[35]

When he finds that the film is not holding together, Antonioni does not look into himself. On the contrary, "I start reconsidering the film, thinking about its features and the way I came to discover them during the preparatory phase on location."[36] While we might call these moments "introspective," they are not products of magnification: they evolve through associative networks of feelings and moods that occurred spontaneously, during the original inspiration.

HIGH ANXIETY

For those who have not recently seen *High Anxiety* (1977), I summarize: Evildoers head nurse Charlotte Diesel (Cloris Leachman) and Dr. Charles Montague (Harvey Korman) have framed Dr. Richard Harpo Thorndyke (Mel Brooks) for a murder committed in a San Francisco hotel. Thorndyke's driver and sidekick, Brophy (Ron Carey), took a picture using a standard 35 mm camera of the murderer, a man disguised as Dr. Thorndyke. That picture, about five by six inches, adorns the front page of the city's newspaper.[37]

Examining the news photo, Thorndyke realizes that he is in a glassed-in elevator thirty feet in the background of the photo, taken at the precise moment of the shooting. He exhorts Victoria Brisbane (Madeline Kahn) to contact Brophy, blow up the negative of the photograph, magnify it many thousands of times, and prove that the tiny image of a man in the elevator is Thorndyke. Brophy does so and we see him make successive enlargements. The final enlargement covers a large wall, at least twenty-five feet square. At that magnification, we can see into the elevator and observe that the small dot has indeed turned into a recognizable image of Dr. Thorndyke: his innocence is proved.

From our godlike perch, we know that Dr. Thorndyke is innocent. But in the real world he would, unfortunately, have no defense. No 35 mm negative— much less a halftone image from a newspaper—can be magnified thousands of times to yield a recognizable photo. Even Nurse Diesel makes this scientific error. She scolds Montague, who has urged them to find the original film and destroy it. She thunders that the "negative" has appeared on the front page of thousands of newspapers in San Francisco. They, she says, can magnify it also. We don't care about this trivial point because the scene and some good jokes depend upon it. (For example, Dr. Montague suggests that to suppress the photo, they buy up all the newspapers in San Francisco. It will cost some $10,000 in dimes.) I suggest that we pursue this joke and learn something about cultural artifacts and the limits of interpretation.

Although Ms. Diesel is the head nurse of the Psychoneurotic Institute for the Very, Very Nervous, she is wrong. The cops cannot enlarge "that negative" because it is a positive print, done in halftones and cannot be enlarged more than three or four times without generating a foggy blur.[38] Her admonition to Dr. Montague and our willingness to accept it denote a common fantasy. This fantasy is that we can always dig deeper into a cultural artifact and find its essences, the secrets that lie just below the surface. By digging deeper or, in this case, by enlarging the artifact into gigantic proportions, we can extract additional information. This was the dominant conceit in *Blow Up*, as it was in an earlier film, *Call Northside 777* (1948), directed by Henry Hathaway. Jimmy Stewart is a tough-talking Chicago newsman, P. J. McNeal, who challenges an eyewitness account in a ten-year-old murder story. He proves the convicted man's innocence by using a blown-up image from a wire service photo—of a newspaper's front page!

This belief in our ability to magnify small bits of a message into ever larger forms animates numerous forms of folk religion. Among them is the apparently universal folk belief in divination. According to true believers, divination is an expert procedure that lets us learn something true by studying an apparently random event. By using esoteric knowledge—gained through extraordinary means either at wizard school or through some occult gift—the diviner can discern signals in apparent randomness. When properly interpreted, this signal will reveal the past or the future. Given the understandable wish to predict the future (or know the truth of an important historical event), we find hundreds of forms of divination.

Common forms of divination are the stuff of everyday life: astrology, palm reading, tarot cards, and the occasional crystal ball appear in popular magazines and daily newspapers. Oneiromancy, or divination by dreams, is also typical of folk psychology. For average persons these forms are less available:

amniomancy—the caul
austromancy—winds
capnomancy—smoke
ceromancy—molten wax dropped into water
ichthyomancy—fish
lampadomancy—the flame of a candle
ophiomancy—snakes
pyromancy—fire

The recipe for creating a form of divination is simple: find the Latin or Greek name of a thing that generates novel patterns and attach the suffix *mancy* to it.

("Mancy" derives from the Old English and Old French word *mancie*, which derives from the Latin word *mantia*, which derives from the Greek verb *manteuesthai*, "to prophesy.")[39]

Better still is a process that we can control without too much difficulty and during which we can say the proper words and ask the proper questions. Thus, we can convert ceromancy into plumbomancy by substituting molten lead for molten wax. In a properly conducted ceremony, we'd ask the spirits to instill their wisdom onto the lead as we pour it into the water. Then, when the lead has solidified into a random form, we can "read" it. If wealthy enough and with sufficient heat, we could use gold instead of lead and thus engage in aurummancy. In most circumstances, the urge to predict the future (or discover what has caused a current malady) is serious: something of great value is a stake, such as the health of a pregnant woman or a child.[40]

Although never fully explained, the notions behind divination include a sense that the forces that produce the seemingly random event, such as the unpredictable shapes produced as the molten lead solidifies, control other seemingly random events, such as sudden illness in a pregnant woman — or misfortune in war or in gambling. By engaging these forces and by using proper means to read them, a skilled person can either cajole them or at least tell us the future. This kind of reasoning depends upon a conviction that from random events we can extract information about complex events in the real world.

This, in turn, depends upon the belief that an information-poor artifact, like a bit of twisted wax or smoke rising from a fire, will generate an information-rich message. To be flat-footed about it, this is impossible since it violates the Second Law of Thermodynamics and its variants in Information Theory: a less-organized structure cannot, by itself, become more organized. Of course, those who affirm ceromancy and its cousins include in their stories of cause and effect God, or some other intelligent being who uses the molten lead to send us an encrypted message. Thus, there is an Organizer behind the apparent randomness.

THE CONVERSATION

Moviegoers will recall the opening scenes of *The Conversation* (1974). We see Harry Caul (Gene Hackman), a middle-aged man in a frumpy rain coat, conduct an elaborate surveillance of a young couple, Mark and Ann, as they wander around a public park filled with a noisy lunchtime crowd. Using multiple microphones placed in multiple locations, Harry and his team record snippets of their conversation. Then, using these recordings and applying electronic masks to them to eliminate nonessential signals, Harry isolates their conversation and amplifies it in his lab.[41] Francis Ford Coppola, the film's writer and director, shot the opening scene using devices—such as shotgun mikes—similar to those we see Harry employ. In this way, the film achieves a kind of verisimilitude: street noises, music, people talking, horns, shouts, and other sounds so flood the scene that we cannot discern the contents of the couple's conversation. We are as ignorant as Harry. When he finally uncovers the conversation, by adjusting audio masks, we are as shocked as he to learn that it could lead to murder. That the identical conversation can sustain two, very different, interpretations is an insight we cannot share until the thrilling conclusion when Harry has to confront his idealized transference to the pretty young woman whose voice he had studied so intently.

Aware of the similar conceit in *Blow Up*, Coppola acknowledged his debt to Michelangelo Antonioni (and to Hitchcock, of course). All these films raise an ethical question: How can technically brilliant experts, one an audio engineer, the other a photographer, having mastered a certain kind of knowledge, control the effects of their discoveries? Both professions, which added together give us cinema, are forms of voyeurism: one is composed of eavesdroppers, the other is composed of peeping toms. If the directors of the two films remained content to record exciting materials about aggression and sex, they would have remained fixed at the level of mere fancy and titillation. However, beyond fancy is lyric truth, and beyond that, high art, which centers upon the dilemmas of veridical representation. Not accidentally, both films turn on this epistemological struggle, and each requires its hero to predict murder, the most archaic of solutions to human conflicts. Coppola recognized that his movie is itself a sublimated instance of voyeurism and that Harry's dilemma is like his—and ours. Harry's dawning awareness that his apparently neutral expertise makes murder more likely is analogous to the discovery by J. Robert Oppenheimer and his colleagues that they had perfected mass murder. That these films lead us to consider these tragic consequences of scientific genius, and bring the mirror up to our faces, marks them as instances of high art.

High art does not mean perfect execution of a total vision. Like most films, the rough cut of *The Conversation* was longer than the finished piece; the opening scene alone, which runs about six minutes in the released version, was shot using six cameras and many microphones. Hence, during the editing process, Coppola and Walter Murch, a master of editing, struggled to select from these numerous shots the visuals and sounds that they felt worked best. Like all artists, they wavered between controlling the materials and being controlled by them.

Taste and maturity are names we give to the ability to find jewels among these accidents. For example, in his screenplay from the mid-1960s, Coppola named his hero Harry Haller, in deference to the hero of Hermann Hesse's novel *Steppenwolf* (1927), a book that Coppola and other rebels of the 1960s found transfixing. *Steppenwolf* narrates the crisis, or disease as Hesse puts it, of a young man who confronts his multiple selves. In doing so, Haller confronts the world of submission and lies created by the stifling middle class against which he seeks to rebel.

From its origins in Coppola's homage to Hesse, the main character's name evolved into Harry Caller, then into Harry Call. By accident, a typist whom Coppola says he was trying to impress, misheard Coppola's dictation and typed "Harry Caul." To not hurt the young woman's feelings and to increase his chances for a date with her, Coppola gave the name more thought and saw how happy this accident was. For a *caul* is another kind of veil, just like Harry's plastic raincoat and the many shots done through translucent screens, shades, and curtains.

Given Coppola's aesthetic rules for the film, the kinds of feelings and mood he wished to evoke in his audience, this name proved ideal.[42] It was no longer an error because he could use it to advance the story. He could trade on its linkage to occult beliefs: that children born with a caul, a portion of the amnion membrane on their heads, were special. (And as I noted above, reading cauls is also a common form of divination.) For example, northern European folktales had it that children born this way could not drown. This belief generated a trade in selling the cauls of newborn to the highest bidder, as we learn in the celebrated opening pages of Charles Dickens's 1850 novel, *David Copperfield*.[43]

We see the allure of folk beliefs in signs and dreams illustrated again in *The Conversation*. While Harry is a superb technician, his talents center upon his ability to extract meaningful sounds from nonmeaningful ones recorded that day in the public square. With these kinds of objects he is supremely confident of his ability to get it right and to accomplish his mission. Like all detectives, including King Oedipus, Harry can be sure that where there is a crime there is a criminal. This basic feature of ordinary causality, that a certain, specific effect has certain, specific causes, lets him piece together the puzzle of the conversation and get it right. Knowing that the two young people are speaking English, Harry can rule out non-English sounds from among those sounds he must retain in the final product. With consummate skill, he reassembles the correct sequence of sounds into English phonemes and thus completes his assignment: to record the objective data emitted by two persons talking at that particular time.

The other, more challenging task, is to interpret the meaning of these words. To do that Harry has to call upon his own associative network, his past and his not-fully-articulated feelings. To graduate from a recorder of sounds to an interpreter of speech, Harry must connect their words to their lived experience. Harry must leave his laboratory and explore the world outside it. But, as we learn, this makes him vulnerable to paranoid reasoning, and it drives him to make possible the thing he fears most. In one scene Harry must hover near a toilet bowl.[44]

Harry's problems begin when he attempts to understand the meaning of the conversation—when he attempts to interpret the speakers' intentions and their lived worlds. Breaking the rules he laid down to his assistant, Stan, Harry asks *why* these people are saying these things and what they mean. Having asked and, he hopes, answered those questions, he can then say something about their future. Like a bumbling anthropologist, Harry stumbles around a world he does not know. To these interpretive tasks Harry brings precisely the worst credentials: a troubled past and a paranoid style. His paranoid style is evident early in the film when Stan defends his curiosity about the couple, claiming that his curiosity is normal: "It's just god-damned human nature." To this Harry responds: "Listen, if there's one sure-fire rule that I have learned in this business it is that I don't know anything about human nature. I don't know anything about curiosity. That's not part of what I do. This is my business."[45] In response to his errors with other people, Harry retreats into smaller and smaller zones of privacy and what he hopes is security. In doing so he decreases access to new information, he impairs his reality testing, and he makes more likely the horror he wished to prevent.

Remaining aligned with its artistic trajectory, Coppola's story deals with the topic of eavesdropping and paranoia at multiple levels (we recall that the Watergate scandal was brewing when he began shooting the script in 1973). To account for Harry's grim life and his fatal mistake, Coppola offers a plausible story of boyhood terror and psychopathology. In a dream, Harry addresses Ann, the young woman he fears will be killed when her husband discovers her love affair, and he recounts his boyhood terror when his mother accidentally left him to nearly drown in the bath tub:

> I could feel the water starting to come up to my chin, to my nose, and when I woke up, my body was all greasy from the holy oil she put on my body. And I remember being disappointed that I survived. When I was five, my father introduced me to a friend of his, and for no reason at all, I hit him right in the stomach with all my strength. And he died a year later. *He'll kill ya if he gets the chance.* I'm not afraid of death but I am afraid of murder.

In this beautiful condensation, Coppola lays out a story that begins with misery, paralysis, suicidal wishes, and miraculous rescue—the holy oil brought the boy back to life—and culminates in inexpugnable guilt. As a boy he punched a man who later died. Harry feels as responsible for that death as he felt responsible for the three persons murdered in a previous case to which he had made an essential contribution.

These emotional struggles have created a man who applies himself relentlessly to the task of hearing secrets but who misunderstands what is under his nose. We see this in a quiet scene with his girlfriend, Amy (played by a young Teri Garr). Amy wants to know more about Harry, to share his inner life. Amy's innocent questions strike Harry as sinister. They validate his decision to keep her disconnected from his real life:

> *Harry:* I don't have any secrets.
>
> *Amy:* (She points to herself) *I'm* your secret. You do have secrets, Harry. I know you do. Sometimes you come over here and you don't tell me. Once I saw you up by the staircase, hiding and watching for a whole hour. (Harry appears embarrassed.) You think you're gonna catch me at something, you know? I know when you come over. I can always tell. You have a certain way of opening up the door. You know, first the key goes in real quiet, and then the door comes open real fast, just like you think you're gonna catch me at something. Sometimes I even think you're listening to me when I'm talking on the telephone.

Amy realizes that Harry fears her, even as he loves her, and that her wish to get closer to him angers him:

> *Amy:* Where do you work, Harry?
>
> *Harry:* Oh, in different places, different jobs. I'm kinda a musician. A free-lance musician.
>
> *Amy:* Where do you live? And why can't I call you over there?
>
> *Harry:* (lying) 'Cause I don't have a telephone.
>
> *Amy:* Do you live alone?
>
> *Harry:* Why are you asking me all these questions?
>
> *Amy:* Because it's your birthday.
>
> *Harry:* I don't want people to ask me a lot of questions.
>
> *Amy:* I want to know you.
>
> Harry: Yes, I know you want to. I don't feel like answering any more questions. [The affair is ended on this sad note as Harry, still clad in his plastic raincoat, gets up angrily:]
>
> *Harry:* You never used to ask a lot of questions.
>
> *Amy:* Harry, I was so happy when you came over tonight. When I heard you open up the door, my toes were dancing under the covers. But I don't think I'm gonna wait for you anymore.

By withdrawing from Amy, a genuine and loving person, Harry falls back into his boyhood notions of adult functioning. These notions, based upon fantasy and not actual encounters, cannot serve him well in adult life. Rather than live with Amy, he seeks comfort in his walled-up apartment. There, sitting alone in a straight-backed chair, he plays his saxophone to accompany a record playing on his phonograph. Harry forsakes a real relationship with Amy to return to a pretend, solipsistic relationship with a dead thing, the recorded sounds coming from yet another mechanical device. The jazz recording was once alive; now it is dead.

As we see many times, Harry misjudges those around him. For example, he admires Ann when he hears her sentimental comments about an old man she spies lying on a park bench: "I always think that he was once somebody's baby boy . . . and he had a mother and a father who loved him. And now, there he is, half-dead on a park bench." (Ann is also wrong since the old man on the park bench is *not* a half-dead drunk, he is one of Harry's operatives secretly taping her conversation.)

In his workspace, Harry hosts a party of bugging experts, among them Moran, his chief competitor. He cannot see that Moran, his slimy archrival, treats him with admiration and contempt. Moran has invited Meredith, his sexy assistant and a good-time girl, to join the party. Still wounded by his loss of Amy, Harry takes Meredith off to a quiet corner and asks her how she would respond to his admonitions to Amy to wait for him and ask no questions:

Meredith:	Something is on your mind. I wish you'd tell me. I do, I wish that, I wish that you'd feel that you could talk to me and, and that we could be friends, I mean, aside from all of this junk.
Harry:	(after a long pause) If you were a girl who waited for someone . . .
Meredith:	You can trust me.
Harry:	. . . and you never really knew when he was gonna come to see you. You just lived in a room alone and you knew nothing about him. And if you loved him and were patient with him, and even though he didn't dare ever tell you anything about himself personally, even though he may have loved you, would you . . . would you, would you go back to him?
Meredith:	How would I know—how would I know that he loved me?
Harry:	You'd have no way of knowing.

To Meredith's reasonable question, Harry has no answer. He admits that neither Amy nor Meredith (nor Stan) can know Harry's true feelings about them because he refuses to divulge anything that would reveal his inner experiences. This leaves them clueless and, eventually, disheartened. Within the logic of his paranoid world, first seen in his angry response to a birthday greeting from his attractive landlady, Harry's effort to share turns into humiliation. Using the gift pen (microphone) he had given Harry earlier, Moran plays a tape recording he has made of Harry's heartfelt talk with Meredith. This shameful moment confirms Harry's worst fears. Not having trusted the genuine woman, Amy, Harry leaves himself open to the false woman, Meredith, who betrays him twice.

To capture Harry's mistaken reading of the meaning of the young couple's discussion about murder, Murch and Coppola capitalized on an error. During the sound-editing process they discovered that the actor who plays Mark, Frederic Forrest, misspoke his lines. In one take, Mark, the young man whom we see talking with Ann, says about Ann's husband, *"He'd kill us if he got the chance."* In a second take, the actor altered the stress. Seizing this opportunity, Coppola and Murch used both takes. Early in the film, Harry hears the first take and deduces that the young people are in danger from Ann's husband. To prevent a recurrence of a tragedy that occurred in New York when his tapes led to the murder of three people, Harry inserts himself into their lives. Later in the film, having realized his error, Harry remembers the line and hears (as do we) the second take.

Coppola shows Harry's drive to undo his past sins by delving into what seems to be a deep secret, something hidden *within* the tapes. Harry lives by the illusion of infinite depth. This illusion—and his compulsion to make up for his errors—prevents him from hearing what is on the surface, from hearing the expression that predicted the murder. Harry's genius was to uncover signals disguised by ambient noise; his error was to make the wrong interpretation. We find a distinguished scientist making the same error.

ISOLATING VALID SIGNALS, MAKING THE RIGHT CUT

Jack Greenberg, a distinguished Yale physicist, illustrates the ease with which even well-trained scientists can share the illusion of infinite depth. Like Harry Caul, Greenberg's story is all about making the right cuts. Greenberg and a group of esteemed colleagues believed that they had discovered a natural phenomenon, a state of matter, never seen before. Using accelerators that smashed heavy elements together, Greenberg and others in the mid-1970s seemed to find evidence for an exotic particle. If correct, Greenberg would be lionized and headed for a Nobel Prize.[46]

The task facing Greenberg and others who took part in the multiyear, multimillion dollar series of experiments was to link up a plausible theory of the elusive, new particle with reliable evidence for its existence. The plausible theory, according to Gary Taubes, emerged in 1969 when a German theorist, Walter Greiner, noted that when two heavy nuclei were smashed together—something that was just beginning to be possible with new accelerators—they might touch momentarily and in so doing create a "quasi-atom," an entity with enormous electrical charge. This giant electric charge, might, in turn, create new, never-before-seen physical phenomena. Greiner speculated that the quasi-atom should emit a spontaneous positron, a positively charged particle at a very high energy for which there was no mundane explanation: "Positrons created by routine physics would have a wide range of possible energies, resulting in a smooth spectrum. But *spontaneous positron emission* or some other exotic process might favor a single characteristic energy, creating a peak or hump in the spectrum" (emphasis added).

In other words, to prove that they had discovered a new form of matter, Greenberg and his colleagues needed to find evidence for unusually high levels of energy displayed against an otherwise expectable spectrum of energy readings. This meant that they had to prove, statistically, that their observations were not mere chance; they had to show that unusual readings could be reproduced reliably. At first, much excitement reigned in Greenberg's group and another one, both running experiments at the Organization for Heavy-Ion Research in Darmstadt, Germany. Both groups found unusual peaks; the trouble was the peaks differed between the two groups, and it proved difficult to reproduce the same findings in subsequent runs of the experiment.

At the same time, a new idea entered into the thinking of the period: perhaps the data represented positrons that were themselves decaying remnants from another object: "In the huge electrical field of the quasi-atom, a new and hitherto unknown neutral particle was created that then decayed into a positron and an electron." This new idea raised the possibility of discovering the new, unknown particle—and that would count as a major coup. The trouble was that it proved difficult to hunt down and capture signs of this new particle. Among those signs would be an electron emitted when the new particle decayed. Retro-fitting their two experiments, both groups now looked for signs that such an electron existed.

Here began an agonizing cycle of apparent success followed by disappointment: in one run of experiments, for example, Greenberg and his group found evidence of an electron associated with a reading of 760 KeV.[47] Since this was highly unusual and seemed to fulfill the requirement of statistical rigor, hopes ran high that they had discovered the elusive signal. Frustration replaced elation, though, when they could not observe the 760-KeV peak again: "These peaks seemed to be reproducible, but in a typically unsatisfying way. They would come and go for no apparent reason, a behavior that Greenberg and his colleagues attributed to subtle changes in the energy of colliding nuclei."

Given the prestige of the scientists involved and plausibility of their findings, Greenberg and others secured additional funding to carry out yet more experiments designed to investigate their original findings of the mysterious peaks. In one experiment, ATLAS Positron Experiment or APEX, researchers spent $2.3 million to investigate the elusive peaks. After running for two years, APEX physicists published their findings: they were unable to replicate Greenberg's findings. Furious, Greenberg and his allies attacked the authors of the paper and sought additional funding to run yet more experiments, which, they said, would show the existence of the mysterious peaks and therefore evidence for the new form of matter. In addition, newly revised and enhanced experiments in Germany, following all of Greenberg's criteria, failed to find any unusual peaks in their new data. Where had the data gone? Or, to put it another way, where had the original data come from?

In a challenge to Greenberg's claims, Rudi Ganz, another physicist, took data from an experimental run and assigned to each of the collisions a random number between 0 and 1. From this single set of data he made two statistically identical sets: "'Everything above 0.5 is data set one,' he says, 'and everything below that is data set two.'" Then, using cuts like those Greenberg employed in his experiments, Ganz applied them to data set one. Among these cuts were adjustments to the time-of-flight of the electrons from the collision to their detection. Fiddling with various cuts, and using high-speed computers, Ganz could detect interesting signals every so often: "Because you really have no idea what you're looking for, you have the freedom to choose any time of flight cuts." Doing so, he was able to find a huge peak at, it happened, 655 KeV.

According to the original research group, this kind of peak was significant and should count as evidence for the existence of the elusive missing particle. When Ganz applied the same cuts and masks to the statistically identical data from data set two he found nothing. This suggests strongly that the huge peak noticed at 655 KeV in data set one was a fluke, based on the massaging of a random event that did not appear in the second set of randomized collision data in data set two. Following Ganz's critique, funding and support for Greenberg's project disappeared.

As we noted in chapter 5, an equivalent error occurs in ESP research. Because a particular subject has guessed twenty out of twenty-five cards correctly and since this is greater than the expected rate of five out of twenty-five (there are five face cards), J. B. Rhine argued that it *proves* the existence of extraordinary powers. From there it is a short distance to claiming that if one person has ESP, then with the right kind of training many could acquire this capacity. Of course, since this is a statistical claim, it is valid only if made within a statistically valid field, that is, using trial runs that are recorded faithfully.

When Irving Langmuir investigated Rhine's studies of ESP he discovered that Rhine retained data that showed a group getting scores above the expected average of five per twenty-five and rejected data from groups that reported scores below the expected average of five per twenty-five. One can imagine Rhine arguing that since we know that the average must be five per twenty-five, trial runs showing less than that average must be aberrant. Hence, Rhine deleted these unattractive results from the totals. Rhine could make these cuts only because his ideology—that ESP was real—justified his bias against results that were "too low." By throwing out results that fell below the expected average of five per twenty-five, Rhine created the evidence that he hoped would convince skeptical critics.[48]

Although Greenberg's story is similar to J. B. Rhine's and other instances of "pathological science," I reject the moralizing tone of that label. For while Rhine seems to verge on outright fraud, Greenberg and the scientists described in the previous chapters do not. Greenberg and his colleagues are, by all accounts, sincere and brilliant people. What went wrong? Nothing went wrong inside of them—there was no upsurge in deviancy, no surplus of unregulated ambition. On the contrary, they erred in performing too few rigorous tests of their exciting ideas. When Rudi Ganz carried out his clever experiment on data set one and data set two, he took part in normal science. His success was Greenberg's failure, true; but it benefited the entire physics community. That is an instance of normal progress in normal science.

If Greenberg and his colleagues had not shared the data sets, Ganz could not have proved them wrong. The rule of normal science—the rudimentary demand that Greenberg show how the experiment was run, for example—is an ethical demand to comply with the self-understanding of the scientific community. This means that the community, not individuals, judges the adequacy of a theory. The community rejects all claims for special privilege and wisdom of individuals. A less obvious consequence of this rule is that no school or group can claim a status invulnerable to criticism and correction. There is no hierarchy in which a bishop or another wise person declares what is acceptable and what is not. An unhappy feature of many humanists is our wish to find such a person and to declare him or her an authority who shall judge the value of new ideas before they are tested. Speaking from within my discipline, psychoanalysis, I know how powerful this wish can be.

MAGNIFICATION IN HUMANISTIC THEORY

That we yearn for such persons and that religions typically elevate some people to positions of immense authority are important facts about human psychology. Because this yearning appears rooted in the child's needs for guidance and security, it reappears in each generation. As the Beatles' song put it, "Mother

Mary comes to me. Speaking words of wisdom, let it be." This is a lovely song. It is a kind of lyric poetry. It shows us yearning within the consciousness of another human being, and that is comforting. The song is its own justification; it works as long as we feel connected to the singer. Like numerous ritual performances, we must repeat the song to retain those feelings. As moving as many rituals are, they are effervescent; they fade and must be renewed permanently.

Each renewal is a link to the idealized past. Each repetition ties us to our parents, and to their parents, all the way back to that single event in which our faith tells us we became truly ourselves. For Heidegger, as we have seen, this idealized past began in Greece around the seventh century B.C.E. when the pre-Socratic philosophers, he says, first asked about Being. For Christians, the idealized past was every day of the life of Jesus, but especially those days in which he was tried, crucified, died, and raised from the dead—as the liturgy puts it. The Last Supper, in which Jesus fed his disciples, is repeated in each enactment of communion because, according to ritual theory, each communion is, in a transcendental sense, the very same action.

For Christians, to be born on one of these days is to feel a special attachment to the faith and to those transcendent times. The current pope, Joseph Ratzinger, who took the name Benedict XVI, was born in Germany (Bavaria) on Holy Saturday, April 16, 1927. He was baptized that day. According to Catholic teaching, baptismal water and other forms of holy water should be consecrated on Holy Saturday. This means that children baptized on that day are "first." Describing his birth and baptism, Ratzinger says: "To be the first person baptized with the new water was seen as a significant act of Providence. I have always been filled with thanksgiving for having had my life immersed in this way in the Easter Mystery."[49]

It is an attractive story. It casts the pope's life into a form of drama: by special providence, God saw fit to make him born on this holy day, and this prefigured his later consecration to the Catholic Church. Like other forms of ritualized religion, the Catholic Church has evolved numerous ways to structure the calendar, and therefore each part of a human life, from birth to death and beyond death. By grounding ritual actions on the calendar and naming and sacralizing each part of life, the church unites each member to all other members and itself to the Holy Spirit and to God.

This has an undeniable majesty. For the small cost of true confession and acceptance of the church's teaching one becomes located in a sacred realm— and in an institution—that cannot be duplicated. A less obvious cost is that orthodoxy by itself cannot take us beyond the level of lyric poetry and its emotional truths. We see this immediately when we ask why baptism is done with water. The authors of the *Catechism of the Catholic Church* cite the Greek verb *baptizein*, which means to "plunge": "The 'plunge' into the water symbolizes the catechumen's burial into Christ's death, from which he rises up by resurrection with him, as 'a new creature.'"[50]

Given the central role that water plays in human life, it is not difficult to see why theological authorities, like poets, find additional reasons why baptism is so important:

1215 This sacrament is also called "*the washing of regeneration and renewal by the Holy Spirit,*" for it signifies and actually brings about the birth of water and the Spirit without which no one "can enter the kingdom of God."

1216 "This bath is called *enlightenment,* because those who receive this [catechetical] instruction are enlightened in their understanding. . . ." Having received in Baptism the Word, "the true light that enlightens every man," the person baptized has been "enlightened," he becomes a "son of light," indeed, he becomes "light" himself.

While these are interesting associations to the use of water, they are poeticisms; they work by evoking symbolic linkages between water and its numerous uses in human lives. In the hands of poetic or theological genius, they can be moving and lyric. Yet, they cannot rise beyond that. Symbolic links are potentially infinite in extension. A contending group of theologians can suggest its own reading of the nature of baptism. If their reading is too distant from the teaching authority, they will find themselves exiled by that authority. To maintain a single, correct reading of its symbols, the church must impose discipline—the task that Professor Ratzinger assumed when he became Pope John Paul II's chief doctrinal officer, then pope. The church offers, in this sense and under these conditions, permanency.

Protestant churches, Islamic clerics, Buddhist masters, and other religious authorities offer similar security by unifying and fixing the metaphorical riches of their traditions. In traditional cultures isolated from competitors, every part of life occurs within the symbolic realm. For example, the authors of *Sakuteiki,* a Japanese treatise on gardening written around 900 C.E. during the Heian period, meld agricultural principles, rules of construction, a theory of nature, and Buddhist ethical teachings into their prescriptions for building a proper garden. Added to these are folk beliefs and taboos that "so heavily influenced every aspect of aristocratic life in the Heian period."[51]

Benedict XVI is an articulate and learned man; he has spent his adult life thinking about the church and its role in the world. In favor of his kind of discipline is the deep truth that like language, spirituality stems from a living tradition. We speak French or English or Chinese, not Language. While many people assert that they are spiritual and not religious, this seems vague and sentimental. If we attach ourselves to the church and enjoy its protection, how should we acknowledge the five billion human beings who remain outside of it? Within the Roman Catholic Church there is a simple answer: we should convert them to our singular truth. This solution seems equivalent to saying that there is no language other than mine and that eventually everyone should speak English in its American idiom.

A humanistic answer might be to recall the distinction between high art and sentimentality. High art lets us enjoy the gifts of metaphor and, at the same time, understand that it remains metaphor. Ruskin said that Dante used the metaphor of falling leaves to describe souls descending even as he distinguished leaves from souls. James Joyce did something similar in his short story "The Dead," when he described all of Ireland, its living and dead, united, secure, and protected under the same blanket of snow.

To affirm this distinction is to resist that part of us that wishes to reduce ambiguities and doubt to singularity and faith. In ideal circumstances, such as Jewish self-reflection, Buddhist psychology, and Christian theology, religious authorities name the gap that exists between our wish to believe and our need to discern intellectual truths. John Donne (1572–1631), a renown Christian poet and intellectual, put it this way: "My God, my God, thou art a direct God, may I not say a literal God, a God that wouldst be understood literally and according to the plain sense of all that thou sayest? but thou art also (Lord I intend it to thy glory, and let no profane misinterpreter abuse it to thy diminuition), thou art a figurative, a metaphorical God too."[52]

Drawing upon his poetic expertise, Donne notes that even brilliant and devoted readers of scripture, such as Saint Augustine, cannot agree upon the meaning of these passages. These brilliant readers cannot agree upon a single, unchanging sense of scripture because scripture turns upon an armature of figurative language. Donne continues his address to God: "The style of thy works, the phrase of thine actions, is metaphorical. The institution of thy whole worship in the old law was a continual allegory: types and figures overspread all, and figures flowed into figures, and poured themselves out into farther figures."[53] Rather than remain plain and direct, even Jesus uses metaphorical language: "How often, how much more often, doth thy Son call himself a way, and a light, and a gate, and a vine, and bread, than the Son of God, or of man?" To illustrate this quality of scripture Donne notes how often scripture speaks about human suffering using water metaphors. He interweaves citations from Hebrew and Christian texts about water and ships with his meditation upon God's promise of safety, the ship "Thine ark, thy ship" out of which "thy blessed Son preached."

In this stream of metaphors we feel the sweep of Donne's understanding of Jesus. Jesus is the ship upon which we sinners, about to drown in error, find rescue. A few paragraphs earlier Donne associated Jesus with water: "All our waters shall run into Jordan, and thy servants passed Jordan dry foot; they shall run into the red sea (the sea of thy Son's blood), and the red sea, that red sea, drowns none of thine."[54] These new metaphors seem to undo Donne's previous metaphors about Jesus. Jesus is the living waters, the sea to whom all rivers flow, the original font of creation, and the sea of blood given that others might live. Jesus is *also* the ship that sails upon the sea. To step away from Donne and his tradition and stand as an outsider to this discourse, we note that when added together these metaphors form a contradiction. Donne said as much in his lament that God is figurative and literal, poetic and historic, to be addressed only though the lens of parabolic reflection: "O, what words but thine can express the inexpressible texture and composition of thy word?"[55]

I find this lovely; but it offers no grounds to consider how progress might occur. Progress means giving up some part of our current beliefs or practices in favor of something new, something that forces us to leave behind a cherished belief. Paradox affirms both that something is true and that it is not true. We affirm both propositions and therefore sacrifice nothing: if Christian teaching requires that we affirm both that Jesus is All Human and All God, then we affirm both propositions and add them to the mysteries of the faith.

Linked to this tolerance for contradictions is idealization of the original generation, for they must have understood what we cannot. Many religious traditions and many philosophical schools assert that somebody before us, the Ancestors, got it just right. As I see it, this wish is another variation of union and permanency. Heidegger took us back to the Greeks, at least to the Greeks as he understood them; Heideggerians take us back to Heidegger. In my profession of psychoanalysis, analysts have a difficult time not taking us back to Freud. In a recent, spirited debate about the future of psychoanalysis—and its scientific status—André Green, a renown French author, argues against reducing psychoanalysis to a common science of behavior.

Contrary to this goal, which is the standard value of most North American academic psychologists, including his opponent, Robert Wallerstein, Green asserts that psychoanalysis is neither science nor humanities. It is a third, independent form of research and theory discovered by Freud. Freud's thought and Freud as the Path Breaker remains underappreciated and understudied. Green states that "I maintain that as yet there is no serious study of Freudian thought in existence today by psychoanalysts."[56] As I noted earlier, Freud remains the most cited author in all of psychology. At least four generations of scholars and psychoanalysts have reflected upon every nuance of every paragraph. In a recent collection of articles and books on psychoanalysis, we find 1,981 citations to the search terms *Freud and science*.[57]

Green's statement might be valid. Perhaps hundreds of readers (psychoanalysts among them) have missed something essential in Freud that a future, better reader will discover. If Green is correct, then these dedicated readers, many of them persons of great learning, have missed an essence, a deep truth in Freud. Where is this truth buried? For that matter, why did Freud in his twenty-four volumes or his many disciples in their voluminous writings not announce it clearly? Ernest Jones, Freud's colleague, confidante, and official biographer, wrote three lengthy books on Freud's life and thought. Jones believed that he grasped Freud's wisdom and importance and compared Freud to the ancient Greeks, who proclaimed "Know Thyself!"—the injunction carved over the temple at Delphi.[58]

According to Green, Jones also missed the essence of Freud's thought. How? This question arises out of an error, an error that treats cultural objects—among them Freud's texts—as if they were natural objects that we can examine in ever greater depths, continuously. Green seems to believe that no one—including himself?—has read Freud correctly. Using terms from this chapter, this amounts to saying that no one has applied the right cuts to the Freudian texts. When someone does apply the right cuts it will reveal the truths that lay hidden within. Jack Greenberg asserted the same thing about data from the many experiments designed to investigate his belief that he had discovered a new state of matter. The decisive test of Greenberg's theory came from a mechanical, objective procedure by which Rudi Ganz created two, identical sets of data and then applied the same cuts to each data set.

Ganz was able to find a striking pattern in one data set, a pattern that seemed far beyond chance and that replicated the kind of discovery that Greenberg reported and that Greenberg felt proved that something unusual had occurred. A rough analogy is looking backward at how stock mutual funds did in the previous three months. Often we find that an obscure fund that specializes in oil drilling stocks, for example, had gigantic returns. Should we therefore throw all our life savings into it? No, unless we wish to lose our money. Given that random events can drive short-term stock prices, the odds are that one of the thousands of mutual funds available will coincide with a random event and generate unusually high returns—for three months. Once the effects of that random event have played out, the fund will fall back into obscurity.

The few hundred people who happened to own the fund and who enjoyed the windfall may feel that they have shown special wisdom and should henceforth invest your money alongside theirs. This would be a mistake. The test of stock market wisdom is how well one does over the long haul, say ten to twenty years. Ganz evaluated Greenberg's theory by following a procedure for making cuts using various, loose criteria. Following Greenberg's methods, Ganz assumed that if the ad hoc cuts he made to data in set one generated a very high reading, they should do the same for the identical data in set two. When it did not, Ganz made it difficult to believe that Greenberg had discovered anything real.

By treating Freud's texts as semisacred documents that we must read in special ways, ways not yet carried out, André Green perpetuates a similar error. I think that many of us do the same thing: we share the belief that we can see deeper into texts and eventually find there hidden truths. Contrary to this belief, I have argued that this model leads us astray. After magnifying a text two or three times, through all the means of reading available, there comes a point where we have nothing more to dissect. Beyond these levels of reading and magnification, the noise level rises to such a state that the text has now become a vehicle for our projections. These projections may be benign, as they are in Donne's reading of the New Testament. Or they may be malign, or self-interested, or destructive as they often are in metaphysical battles over esoteric meanings. It is true that a brilliant reader can "find" in Freud startling new things to say, and every so often, those new things may be true. Yet, we have no easy way to distinguish what is true from what is merely clever.

This brings us back to the unhappy fact that often humanists and psychoanalysts coagulate into schools, each with its special truth, its private language, and its hotly contested boundaries. This is depressing in a special way. Donne acknowledges that his struggle with scripture derives from an emotional conflict: his heart says that scripture is true while his reason tells him that we find within it contradictions, and contradictions are always false. Donne's way out of this conflict is to rename the contradiction and call it a paradox. How can humanists, and psychoanalysts in their humanist modes, find ways to address *their* conflicts?

Concerning this issue, Green announces that psychoanalysis is neither humanities nor science, but a third thing: "It is a practice based on *clinical thinking* that leads to theoretical hypotheses."[59] I would like to believe this. Having cast my lot with psychoanalysis and the humanities, I hope that my work begins with clinical thinking and can generate hypotheses. While this does not make psychoanalysis a natural science, these terms describe a type of empirical study. We listen to patients; we observe the intense relationship of therapy; we look for patterns in beliefs, feelings, and actions; and we form hypotheses about these patterns. We compare these claims of pattern against new evidence—in this sense we test our hypotheses.

This sounds attractive and reasonable. It is also, I would note, not an exercise in deep reading. In the metaphor of vertical and horizontal, clinical generalizations are only one or two levels deep. For example, a woman who is usually on time and cheerful in the morning arrives one morning late and angry. We observe this change in pattern: we wonder why it occurred; we listen to the patient's free associations; we observe her behavior; we listen for derivatives, that is, displaced comments about us and perhaps an error we made in the previous session. Different clinical styles and theories may make us focus on one thing rather than another. In all these versions, however, we do not stray very far away from the patient's manifest material.

When analysts or other therapists look for deep meanings, they must abandon these descriptive levels and with them the patient's immediate experience. Deeper than mere self-awareness are the actions of agencies that theory tells us must be there. Theory, in this case, is nothing more than imagination for having left the surface (or the "mere" surface), and perhaps after two or three levels of magnification, we have no landmarks. For example, Freud's notion of the "Death Instinct" and "Eros" is a dramatic story, a mythology, a brilliant metaphor, perhaps. It links Freud's thought back to the Greeks, but it cannot help us generate testable hypotheses.

Like other humanist thinkers, or, as he prefers, psychoanalytic thinkers, Green believes that theory shapes data to such an extent that there is no common ground to analysts who hold different theories. This means that when Wallerstein and others seek consensus, they are in error: "*The only valid procedure is to show how some material consisting in, thus based on, the exposition of a sequence of sessions and on a psychoanalytic process revealed at sufficient length can demonstrate the kinship between two different theories, which we must remember are based on different techniques and interpretations.*"[60] Green adds that he has never seen this done. On the contrary, "The adherents of one theory are often opposed to another on every possible count: the clinical understanding of the material, the technique adopted and the theory that explains them."[61] To these American ears, this is all unfortunate, as are Green's comments that he fears researchers who cite texts from observational scientists. These texts and studies introduce, he says, "anti-psychoanalytic points of view" that are analogous to viruses that will harm the core of psychoanalysis. When Freud was alive we had someone who could declare what counted as psychoanalytic and thereby preserve the integrity of the discipline, Green says.

This is a lucid description of a fundamental error. The error is one of circular reasoning that plagues many discussions within humanistic and psychoanalytic groups. It goes like this: (1) there is a core or essence of psychoanalysis (or art history, etc.) that some special persons have understood and many have not; (2) those who know this truth can communicate it to their disciples — perhaps — but only time will tell if the disciples are true or false; (3) other, competing ideas are dangerous, like viruses, because they distort the essential truth and take us away from the master's teachings; (4) other ideas and those who carry them must be isolated or excluded from the healthy body, else destruction ensues; and (5) proof of the master's teaching is available only to those who grasp it. This is a pernicious circle. Persons who challenge the master, either within or without, reveal by their actions that they are unworthy of membership.

Within the circle of true believers one sometimes finds brilliant defenders of the faith. Augustine emerges as a person of unusual talent, as did Heidegger; certainly Freud belongs to that group as well. When Hannah Arendt repledged herself to Heidegger we see a person of vast talent exchange her enviable skills for obedience. If it can happen to her, it can happen to me and others who do not match her intellectual powers. I have suggested that this yearning for sureness is stronger than an individual's commitment to the truth. Jack Greenberg emerges as a smart man caught up in an all-too-human wish to make his wishes a new truth. Few of us could do otherwise. The only thing that preserved the science of physics was Greenberg's openness to others (and their viruses). The interpersonal values of free exchange, the ethics of science as Richard Feynman put it, required Greenberg to share his complete data with Rudi Ganz.

But this is what humanists—both poets and professors, both artists and art historians—cannot do. We do not have a complete data set. *Our* subject matter, human being, is not fully bounded: to declare limits to human being is to engage in ideology; to claim exhaustive knowledge of human action and human destiny is to make theological, not scientific claims. For this reason humanists cannot escape the task of theology. Even if they are adamant in their atheism, humanists must deal with that part of us that demands an ideal Other against whom we measure ourselves, through whom we imagine new possibilities of human being. For religious persons, this Other has always been named God; for nonreligious humanists, this Other typically is an idealized genius. In the nineteenth century, G. F. W. Hegel occupied this position for many European and American intellectuals. In the twentieth century Nietzsche and Freud took Hegel's place. At the beginning of the twentieth-first century it is not clear which names will rise to those Olympian heights.

— — —

I have used the metaphor of magnification to help distinguish what scientists study from what humanists study. I suggested that because scientists study natural objects, they can magnify, that is dissect into smaller and smaller parts, the things they study. At every new level of magnification new structures appear and therefore new information emerges. There is always more to see. This singular fact means that natural and biological scientists can draw upon insights and "laws" generated by other scientists working at different levels of organization. It does not mean that natural scientists can ignore issues like emergence, complexity, and similar features of the natural world that do not yield to this kind of unidirectional reductionism. Just because scientists affirm that nature is built up of smaller entities, we do not agree that to understand the behavior of complex events—from ocean currents to political movements—we need only reduce them to their parts. On the contrary, many smart people argue that human behavior, like ocean currents, cannot be understood except as partly undetermined and indeterminable.

While the metaphor of magnification is limited, I suggest that it is valuable because it helps us distinguish the task of the sciences from the tasks—and dilemmas—of the humanities. The world of science is dedicated to finding unity in diversity. The goal of scientific inquiry is to discover the simpler, shared characteristics of otherwise different things. Beginning in the sixteenth century through our time, the most elegant instance is physics and physicists' pursuit of a Theory of Everything.

In contrast to science, the world of the high arts is dedicated to novel discoveries even in ancient texts. On first seeing a movie we ask "What comes next?" A second viewing is a different experience. The first viewing of *The Godfather* (1972) tells how Michael Corelone saves his father's life in the hospital. A second and third viewing gives us the pleasure of seeing how Francis Ford Coppola tells the story in a dense, exciting way; how a young Al Pacino expresses anger; how empty the hospital looks; etc. Viewing yet another version of *Hamlet*, we are not surprised by Hamlet's death, we are moved by the quality of a specific performance. The plot of Mozart's opera *The Marriage of Figaro* (1786) remains the same in each new performance. But when done by a good cast the opera captivates us just as ancient religious rituals captivate modern participants.

Art enhances art, and high art enhances life. Peter Shaffer's play *Amadeus* brings Mozart back to us. In one scene Mozart's wife, Constanze, brings a half dozen scores to Antonio Salieri (1750–1825), who will judge their worth before he, as Court Composer, agrees to sponsor Mozart to Emperor Joseph II. The film version shows Constanze and Salieri speaking in 1780 and intercuts with a framing narrative of Old Salieri in 1823 addressing a young priest named Vogler, to whom he has confessed his (delusional) belief that he murdered Mozart. We begin with the scene set in 1780, when Salieri asks Constanze to leave the scores for him to examine at a later time:

> *Constanze:* That's very tempting, but it's impossible, I'm afraid. Wolfi would be frantic if he found those were missing. You see, they're all originals.
> *Salieri:* Originals?
> *Constanze:* Yes.
> A pause. He puts out his hand and takes up the portfolio from the table. He opens it. He looks at the music. He is puzzled.
> *Salieri:* These are originals?
> *Constanze:* Yes, sir. He doesn't make copies.

CUT TO INT. OLD SALIERI'S HOSPITAL ROOM—NIGHT—1823

> The old man faces the Priest.
>
> *Old Salieri*: Astounding! It was actually beyond belief. These were first and only
> drafts of music yet they showed no corrections of any kind. Not one.
> Do you realize what that meant?
> Vogler stares at him.
>
> *Old Salieri*: He'd simply put down music already finished in his head. Page after
> page of it, as if he was just taking dictation. And music finished as no
> music is ever finished.

INT. SALIERI'S SALON: LATE AFTERNOON—1780's
CLOSE UP, The manuscript in Mozart's handwriting. The music begins
to sound under the following:

> *Old Salieri*: (VOICE OVER) Displace one note and there would be diminish-
> ment. Displace one phrase, and the structure would fall. It was clear
> to me. That sound I had heard in the Archbishop's palace had been
> no accident. Here again was the very voice of God! I was staring
> through the cage of those meticulous ink-strokes at an absolute, inimi-
> table beauty.

The music swells. What we now hear is an amazing collage of great passages
from Mozart's music, ravishing to Salieri and to us. The Court Composer,
oblivious to Constanze, who sits happily chewing chestnuts, her mouth covered
in sugar, walks around and around his salon, reading the pages and dropping
them on the floor when he is done with them. We see his agonized and wonder-
ing face: he shudders as if in a rough and tumbling sea; he experiences the point
where beauty and great pain coalesce. More pages fall than he can read, scatter-
ing across the floor in a white cascade, as he circles the room.

Finally, we hear the tremendous Qui Tollis from the Mass in C Minor (*Qui
tollis peccata mundi*, "Thou that takest away the sins of the world"). It seems to
break over him like a wave and, unable to bear any more of it, he slams the
portfolio shut. Instantly, the music breaks off, reverberating in his head. He
stands shaking, staring wildly. Constanze gets up, perplexed.

> *Constanze*: Is it no good?
> A pause.
> *Salieri*: It is miraculous.

Salieri swoons with pleasure when he reads Mozart's scores, and we agree
that the music is miraculous. We feel what we felt when we first discovered

Mozart. Through Salieri we hear the orchestral rendering of 200-year-old music as new because a superb actor takes us into his heightened experience.[62]

Is the movie *Amadeus* an example of progress in the humanities? Yes, it won eight Academy Awards and universal praise. Yet, its writers, director, actors, and musicians did not dissect the life of Mozart, nor did they dissect his music. On the contrary, as historians attest, Peter Schaffer "mythified" Mozart's story and transformed Salieri into a villain.[63] Like the three films discussed earlier, *Amadeus* is valuable because it expands our sense of what it is to be a human being in a certain place and a certain time. Schaffer's insight was to create a Salieri with whom we nongeniuses identify. Of course, rigorous historians cannot grant the movie standing within their discipline: they dissolve its heroic narratives into more exact, more "scientific" descriptions.

We have learned to scorn the rhetoric of triumphalist English historians like Macaulay and American zealots who read the fall of the USSR as a sign of American destiny.[64] We obey these admonishments and we abstain, henceforth, from sharing triumphalist feelings. Yet, I note the persistence of our yearning for a narrative that explains our past—like the demise of the Soviet Union—and predicts, even if obscurely, our future. That we share this yearning with ancient peoples may not, all in all, be a bad thing. To return to *Amadeus*, we recall that Mozart's great Mass in C Minor is a piece of music set to the words of the Catholic Mass first codified around the sixth century. Thus the "medium" level art of a movie depends upon the "high art" of Mozart, which in turn depends upon the nonart of ancient scriptures. For Christians, of course, the correct grading system is the opposite: the "high," if not infinite, value of the Mass derives from its linkages to the Last Supper. That unbroken chain of spiritual power begins with the New Time established by the God/Man. His authority animated Mozart, whose attunement to what is sacred animated Peter Shaffer and us, the audience. As Salieri says, surely this was the voice of God.

We do not need art to dissect our emotions into smaller pieces. Great artists and great humanists—like Freud, James Joyce, and Jane Austen—do not see deeper into us the way that biologists peer deeper into the hidden structure of the cell. We call Joyce and Austen deep as ways to honor them. But this metaphor of depth is, I have suggested, misleading. Austen, Joyce, and Freud are exacting naturalists; they explore the texture of lived experience, of the mere surface. Yet the surface is where we live. We may say that a dream is from the unconscious, but that is just a way of talking: a dream is one among numerous features of intense, conscious experience.

We need art to amplify parts of ourselves into larger, more complete wholes. When successful, artists and humanist critics help make seen what was unseen. We need art (and some need religion) to magnify what is obscure in us into something visible, tangible, and real to all of us. That counts, I suggest, as progress.

NOTES

INTRODUCTION

1. W. Watson & Sons 313 High Holborn London #10251. The Van Heurck No. 1 model microscope c. 1908. Antique Microscopes, http://users.bestweb.net/~wissner/index.html (accessed September 6, 2008).

1. A NEW ANSWER

1. See http://www.vanderbilt.edu/AnS/religious_studies/gay.htm.

2. Ernest Jones, *The Life and Work of Sigmund Freud*, vol. 2: *Years of Maturity, 1901–1919* (New York: Basic Books, 1955), 423.

3. Hannah Arendt, "Martin Heidegger at Eighty," *New York Review of Books* 17/6 (1971).

4. In a footnote, Arendt asserts "but the point of the matter is that Heidegger, like so many other German intellectuals, Nazis and anti-Nazis, of his generation never

read *Mein Kampf*." Hitler's book, *Mein Kampf*, was published in 1926 and quickly became the standard text of National Socialism. High school science books cited it, as did numerous party propaganda pieces. *Mein Kampf* is not complex; a person of Heidegger's talent could grasp its core idea—that noble Germans were betrayed by evil Jews—without devouring it page by page.

5. K. Esser, U. Lüttge, W. Beyschlag, J. Murata, eds., *Progress in Botany*, vol. 65 (New York: Springer, 2004).

6. Alasdair MacIntyre, *After Virtue*, 2nd ed. (South Bend: University of Notre Dame Press, 1984).

7. Java Distributed Computing, http://www.unix.org.ua/orelly/java-ent/dist/ch01_01.htm (accessed September 6, 2008).

8. As of April 17, 2005, SETI reports that more than five million users have donated a total of 2,267,570 CPU years to the project. Seti@Home, http://setiathome.ssl .berkeley.edu/ (accessed September 6, 2008).

9. Or, the universe might be as young at 11.2 billion years. See Lawrence M. Krauss and Brian Chaboyer, "Age Estimates of Globular Clusters in the Milky Way: Constraints on Cosmology," *Science* 299 (2003): 65–69.

10. RHIC, http://www.bnl.gov/rhic/QGP.htm (accessed September 6, 2008).

11. Reinhold Niebuhr, *The Nature and Destiny of Man: A Christian Interpretation* (New York: Scribner, 1943).

12. In "The Power of Conversation: Jewish Women and Their Salons," Greta Berman writes: "Numerous young intellectuals and scientists attended her Berlin gatherings. So important was Madame Herz that she—like Queen Marie-Antoinette, a few years earlier—was depicted in 1778 as the goddess Hebe. This portrayal of a Jewish woman in an allegorical pagan guise reflected the acculturation of Jews into mainstream German culture, known as the Jewish Enlightenment movement, or *haskalah*." *The Juilliard Journal*, http://www.juilliard.edu/update/journal/j_articles499. html (accessed September 6, 2008).

13. Anna Dorothea Therbusch, *Henriette Herz as Hebe*, 1778, oil on canvas, Nationalgalerie, Staatliche Museen zu Berlin-Preussischer Kulturbesitz.

14. Henry David Thoreau, *Walden*, ch. 5: "Solitude." The Free Library, http:// thoreau.thefreelibrary.com/Walden-&-on-the-Duty-of-Civil-Disobedience/1– 5#Aesculapius (accessed September 6, 2008).

15. Supposedly, "the term comes from the Skonk Works, the Kickapoo Joy Juice bootleg brewing operation in Al Capp's *Li'l Abner* comic strip." Webopedia, http:// www.webopedia.com/TERM/S/skunkworks.html (accessed September 6, 2008).

16. "To invent the IBM PC, IBM created three secret research teams who competed against each other. The winner was the research team headed by Philip 'Don' Estridge in Boca Raton, Florida." The History of Computing Project, http://www. thocp.net/biographies/estridge_don.html (accessed September 6, 2008).

17. T. S. Kuhn, *The Structure of Scientific Revolutions* (Chicago: University of Chicago Press, 1962).

18. "In recent speeches and news conferences, Defense Secretary Donald H. Rumsfeld and the nation's senior military officer have spoken of 'a global struggle

against violent extremism' rather than 'the global war on terror,' which had been the catchphrase of choice. Administration officials say that phrase may have outlived its usefulness, because it focused attention solely, and incorrectly, on the military campaign." *New York Times*, http://www.nytimes.com/2005/07/26/politics/26strategy.html?oref=login (accessed September 6, 2008).

19. Ibid.

20. Roy P. Basler, ed., *The Collected Works of Abraham Lincoln* (Fredericksburg: Rutgers University Press, 1953), 3.522.

21. *The Da Vinci Code* Official Homepage, http://www.danbrown.com/novels/davinci_code/plot.html (accessed September 6, 2008).

22. Harold J. Gordon Jr., "Introduction," in *The Hitler Trial before the People's Court in Munich*, trans. H. Francis Freniere, Lucie Karcic, and Philip Fandek (Arlington, VA: University Publications of America, 1976), 1.xix.

23. Nicole Krauss, in the *Boston Review*, http://www.bostonreview.mit.edu/BR25.3/krauss.html (accessed September 6, 2008).

24. Samuel Johnson, "The Plan of the English Dictionary," edited by Jack Lynch: http://andromeda.rutgers.edu/~jlynch/Texts/plan.html (accessed September 6, 2008).

25. In their popularized guise, "quantum effects" designate unusual events where people or objects act in ways contrary to common sense and to Newtonian mechanics, analogous to the ways that elementary particles act under certain conditions. Given the oddity of the latter, it is tempting to associate all kinds of other oddities and mysteries to it. Thus a web page reports a possible "quantum" effect between people: Dr. Peter Fenwick, a clinical neurophysiologist, "found that one half of a couple who have a strong emotional connection—lovers or parent and child—can tell when something is happening to their other half although both are in separate rooms and cannot hear or see each other." Strange Birth of a Brainwave, http://members.fortunecity.com/templarser/bwave.html (accessed September 6, 2008).

26. Klaus Theweleit, *Male Fantasies*, trans. Stephan Conway, Erica Carter, and Chris Turner (Minneapolis: University of Minnesota Press, 1958), 1.19–24.

27. Theweleit, *Male Fantasies*, 23.

28. Gordon, "Introduction," in Hitler Trial, 1.v.

2. MAGNIFICATION AND CULTURAL OBJECTS

1. Hiroshige, *The Sea at Satta in Suruga Province*. Series: *Thirty-six Views of Mt. Fuji*. Medium: woodblock. Signature: Hiroshige ga. Publisher: Koeido (Tsutaya Kichizo). Carver: Horicho. Date: Ansei V/4; 4th month, 1858. Hiroshige at Artline.com, http://www.artline.com/associations/ifpda/ifpdafair/ifpdafair99/exhibitors/staley/hiroshige.jpg (accessed September 6, 2008). Compare this with Paul Cézanne's *Mont Sainte-Victoire au grand pin 1886–1887*. Phillips Collection, Washington, D.C.

Artrecordiff.com, http://www.artrecordiff.com/impression/cezanne/cezanne069.jpg (accessed September 6, 2008).

2. Ken Johnson, "How a Japanese Master Enlightened the West," *New York Times*, July 1, 2005, B31.

3. "Ces mots de Gauguin illustrent bien l'usage si particulier qu'il fait du rose, son amour de plus en plus vif pour l'indigo et le jaune citron, la profondeur de ses ocres rouges, le balancement du vert du suraigu au très grave, ses harmonies sombres, presque sourdes, déchirées par des dissonances." Artrecordiff.com, http://www.artrecordiff.com/impression/cezanne/index.html (accessed September 6, 2008).

4. Michael Kimmelman, "With Music for the Eyes and Colors for the Ear," *New York Times*, July 1, 2005, B33.

5. Richard Wagner, *The Art-Work of the Future: Richard Wagner's Prose Works*, trans. William Ashton Ellis (Ithaca: Cornell University Library, 1895), 1.69–213; idem, *Das Kunstwerk der Zukunft* (Leipzig: Hesse & Becker, 1849); idem, *Sämtliche Schriften und Dichtungen* (Leipzig: Hesse & Becker), 1.194–206. Quotations from the Art-Work of the Future, http://users.belgacom.net/wagnerlibrary/prose/wagartfut.htm (accessed September 6, 2008).

6. "The Jew, who is innately incapable of enouncing himself to us artistically through either his outward appearance or his speech, and least of all through his singing has nevertheless been able in the widest-spread of modern art-varieties, to wit in Music, to reach the rulership of public taste.—To explain to ourselves this phenomenon, let us first consider how it grew possible to the Jew to become a musician." "Judaism in Music," http://users.belgacom.net/wagnerlibrary/prose/wagjuda.htm (accessed September 6, 2008).

7. And: "The Jew—who, as everyone knows, has a God all to himself—in ordinary life strikes us primarily by his outward appearance, which, no matter to what European nationality we belong, has something disagreeably foreign to that nationality: instinctively we wish to have nothing in common with a man who looks like that." "We can conceive no representation of an antique or modern stage-character by a Jew, be it as hero or lover, without feeling instinctively the incongruity of such a notion." "Yet his whole position in our midst never tempts the Jew to so intimate a glimpse into our essence: wherefore, either intentionally (provided he recognises this position of his towards us) or instinctively (if he is incapable of understanding us at all), he merely listens to the barest surface of our art, but not to its life-bestowing inner organism."

8. Roberta Smith, "Body Heat: Mannerism and Mapplethorpe Muscles in Tight Embrace," *New York Times*, July 1, 2005, B36.

9. Abraham Flexner, *Medical Education in the United States and Canada: A Report to the Carnegie Foundation for the Advancement of Teaching*, Carnegie Foundation for the Advancement of Teaching Bulletin 4 (New York: Carnegie Foundation, 1910), 602.

10. Bruce E. Wampold, *The Great Psychotherapy Debate: Models, Methods, and Findings* (Mawah, NJ: Erlbaum, 2001). Thanks to Paul Mosher for citing this text.

11. The relevant text begins with the Dodo Bird laying out a racecourse "in a sort of circle, ('the exact shape doesn't matter,' it said,) and then all the party were placed along the course, here and there. There was no 'One, two, three, and away,' but they

began running when they liked, and left off when they liked, so that it was not easy to know when the race was over. However, when they had been running half an hour or so, and were quite dry again, the Dodo suddenly called out 'The race is over!' and they all crowded round it, panting, and asking, 'But who has won?' This question the Dodo could not answer without a great deal of thought, and it sat for a long time with one finger pressed upon its forehead (the position in which you usually see Shakespeare, in the pictures of him), while the rest waited in silence. At last the Dodo said *'Everybody has won, and all must have prizes.'"* Cleave Books, http://www.ex.ac.uk/trol/grol/alice/won03.htm (accessed September 6, 2008).

12. Wampold, *Great Psychotherapy Debate*, 204.

13. Jerome D. Frank and Julia B. Frank, *Persuasion and Healing: A Comparative Study of Psychotherapy* (Baltimore: Johns Hopkins University Press, 1961), 206.

14. Claude Lévi-Strauss, *Tristes Tropiques: An Anthropological Study of Primitive Societies in Brazil* (New York: Atheneum, 1965); idem, *Totemism*, trans. R. Needham (Boston: Beacon, 1971). See also Volney P. Gay, "Ritual and Psychotherapy: Similarities and Differences," in *Religious and Social Ritual*, ed. V. De Marinis and M. Aune (Albany: State University of New York Press, 1996), 217–34.

15. Volney Gay, *Joy and the Objects of Psychoanalysis* (Albany: State University of New York Press, 2001).

16. It may be that definitions of character pathology are bogus and that the entire field is an error covering up our inability to locate the real cause of psychopathology. This real cause might be an as-yet-unseen neural disorder. In that case, biological reductionism will reveal, in some future time, the hidden pathogens responsible for the manifest disorder. Or, the real cause might be social-cultural structures (an oppressive family, culturally embedded misogyny) that manifest themselves in disordered behaviors that we incorrectly label as personal psychopathology. One can find articulate defenders of both critiques.

17. M. Gardner, "Is That a Fact? Empiricism Revisited, or a Psychoanalyst at Sea," *International Journal of Psychoanalysis* 75 (1994): 92.

18. André Green, "Conceptions of Affect," *International Journal of Psychoanalysis* 58 (1977): 129.

19. Philip Lopate, "With a Girl, a Gun, and a Dreamy Dance: A Gift from Godard," *New York Times*, August 12, 2001.

20. Joyce Carol Oates, "John Updike's *Rabbit at Rest*," *New York Times Book Review*, Sept. 30, 1990; see Celestial Timepiece, http://www.usfca.edu/fac-staff/southerr/rabbit.html (accessed September 6, 2008).

21. Michael Drosnin, *The Bible Code* (New York: Simon & Schuster, 1977). For a summary of refutations, see Robert Todd Carroll, "The Skeptic's Dictionary," http://skepdic.com/bibcode.html (accessed September 6, 2008); Dave Thomas, "New Mexicans for Science and Reason," http://www.nmsr.org/biblecod.htm#jordan (accessed September 6, 2008). For a report on efforts to replicate the discovery of patterns, see http://cs.anu.edu.au/~bdm/dilugim/gans_exp.html (accessed September 6, 2008). See also Brendan McKay, Dror Bar-Natan, Maya Bar-Hillel, and Gil Kalai, "Solving the Bible Code Puzzle," *Statistical Science* 14 (1999): 150–73.

22. These are modest estimates. One scholar asserts that about ninety-six bil-

lion persons have lived since around 10,000 B.C.E. See http://www.math.hawaii.edu/~ramsey/People.html (accessed September 6 2008). Another educated guess is about 105 billion.

23. William James, "Psychical Research," in *The Will to Believe and Other Essays in Popular Philosophy and Human Immortality* (repr. New York: Dover, 1956), 327.

24. Robert G. Jahn and Brenda J. Dunne, "Science of the Subjective," presented at "a symposium held in the John M. Clayton Hall of the University of Delaware on September 27–29, 1997, entitled 'Return to the Source: Rediscovering Lost Knowledge and Ancient Wisdom,' which was sponsored by the Society for Scientific Exploration and supported in part by a generous grant from the Lifebridge Foundation." Objective vs. Subjective Information, http://www.scientificexploration.org/jse/articles/jahn1/2.html (accessed September 6, 2008).

25. Henri Bergson, *An Introduction to Metaphysics*, 2nd ed., trans. T. E. Hulme (Indianapolis: Bobbs-Merrill/New York: Liberal Arts Press, 1955), 53–54.

26. Meehl cites a "congress" at the resort town of Achensee in the Austrian mountains, at which Freud and Wilhelm Fliess, his intensely important friend and confidante, discussed the validity of Freud's interpretations of his patients' dreams. Fliess asked Freud if he were not merely projecting his own thoughts into his patients' minds. This possibility, that psychoanalytic interpretations are merely clever poeticizing, galvanized Freud and subsequent generations of psychoanalysts. See Paul Meehl, "Subjectivity in Psychoanalytic Inference: The Nagging Persistence of Wilhelm Fliess's Achensee Question," *Psychoanalysis and Contemporary Thought* 17 (1994): 3–82. See also, Donald P. Spence, "When Do Interpretations Make a Difference? A Partial Answer to Fliess's Achensee Question," *Journal of the American Psychoanalytic Association* 43/3 (1995): 689–712.

27. Meehl, "Subjectivity in Psychoanalytic Inference," 65–75.

28. Meehl, "Subjectivity in Psychoanalytic Inference," 75.

29. Meehl, "Subjectivity in Psychoanalytic Inference," 78. This strategy has an exact parallel in Lévi-Strauss's famous recipe for "reading a myth." See Claude Lévi-Strauss, "The Structural Study of Myth," *Journal of American Folklore* 67 (1955): 428–44, reprinted in *Reader in Comparative Religion*, ed. W. A. Lessa and E. Z. Vogt (New York: Harper & Row).

30. Sigmund Freud, "Notes upon a Case of Obsessional Neurosis" (1909), in Standard Edition 10.155. Unless otherwise noted, all Freud references are to *The Standard Edition of the Complete Psychological Works of Sigmund Freud*, ed. James Strachey, 24 vols. (London: Hogarth/Institute for Psycho-Analysis), cited by volume and initial page number.

31. See Volney P. Gay, *Understanding the Occult* (Minneapolis: Fortress, 1989).

32. Institute for Creation Research, http://www.icr.edu (accessed September 6, 2008).

33. Among members in 1885 were Professor G. Stanley Hall (Johns Hopkins University), Professor George S. Fullerton (University of Pennsylvania), Professor Edward C. Pickering (Harvard College Observatory), Dr. Henry P. Bowditch (Harvard Medical School), and Dr. Charles S. Minot (Harvard Medical School).

34. William James, "The Final Impressions of a Psychical Researcher," *American Magazine*, October 1909, reprinted in *William James on Psychical Research*, ed. Gardner Murphy and Robert O. Ballou (London: Chatto & Windus, 1961).

3. BACK TO FREUD, BACK TO THE GREEKS!

1. Christopher Alexander, *The Timeless Way of Building* (New York: Oxford University Press, 1979).

2. "It would be fair to say that there are few twentieth century thinkers who have had such a far-reaching influence on subsequent intellectual life in the humanities as Jacques Lacan. Lacan's 'return to the meaning of Freud' . . . profoundly changed the institutional face of the psychoanalytic movement internationally." Matthew Sharpe, University of Melbourne. The Internet Encyclopedia of Philosophy, http://www.iep. utm.edu/l/lacweb.htm (accessed September 6, 2008).

3. Joseph Campbell, ed., *The Portable Jung* (New York: Viking, 1971), 55.

4. The concept of *eid lon* is too tempting to not submit to occult processing. Thus a website tells us: "An eidolon (from Greek *eidolon* form, shape; a phantom-double of the human form; Latin *simulacrum*) is the astral double of living beings; the shade or *perisprit*, the *kama-rupa* after death before its disintegration. The phantom which can appear under certain conditions to survivors of the deceased." Quoted from Answers. com, http://www.answers.com/topic/eidolon (accessed September 6, 2008).

5. All citations from Ernest Jones, *The Life and Work of Sigmund Freud*, vol. 2: *Years of Maturity, 1901–1919* (New York: Basic Books, 1955), 423.

6. Martin Heidegger and Eugen Fink, *Heraclitus Seminar 1966/67* (1970), 162.

7. Martin Heidegger, "The Fieldpath," trans. Berrit Mexia, *Journal of Chinese Philosophy* 13 (1986): 455–58 (original 1948); http://www.omalpha.com/jardin/heidegger1-eng-imp.html (accessed September 6, 2008). See also Ereignis, http://www.webcom.com/~paf/href.html (accessed September 6, 2008).

8. Martin Heidegger, "The Self-Assertion of the German University," trans. William S. Lewis, *Review of Metaphysics* 38 (1985): 467–81 (original 1933).

9. Heidegger, "Self-Assertion of the German University," 471–72.

10. Heidegger, "Self-Assertion of the German University," 472.

11. "For Nietzsche, as for Heidegger, European history—and that means, for both thinkers, essentially German history—merely repeats, amplifies, and fulfills the Greek model." Glen Most, "Heidegger's Greeks," *Arion* 10.1 (Spring/Summer 2002): 83–98.

12. Wilhelm Conrad Roentgen won the first Nobel Prize in physics in 1901; Max Planck won it in 1918; Johannes Stark in 1919; Albert Einstein in 1921; James Franck and Gustav Hertz in 1925. A year before Heidegger spoke, Werner Heisenberg won the Nobel Prize for physics; in October 1933, four months after Heidegger's address, Erwin Schrödinger won the physics prize.

13. Heidegger, "Self-Assertion of the German University," 473.

14. Heidegger, "Self-Assertion of the German University," 473–74.

15. Martin Heidegger, "Hegel and the Greeks," from Conference of the Academy of Sciences at Heidelberg, July 26, 1958. http://www.morec.com/hegelgre.htm (accessed September 6, 2008).

16. T. S. Kuhn, *The Structure of Scientific Revolutions* (Chicago: University of Chicago Press, 1969); Karl Popper, *Conjectures and Refutations: The Growth of Scientific Knowledge* (New York: Harper, 1963).

17. All *Phaedrus* citations from Benjamin Jowett's translation at http://www.ac-nice.fr/philo/textes/Plato-Works/20-Phaedrus.htm (accessed September 6, 2008).

18. "Now, the message of this first part is clear, though it is summarized cryptically in a formula already mentioned at the beginning of this presentation, and found at the start of the second subsection of the third section (263d): there is no way a *logos* can be meaningful unless it takes into account the *whole* of man, body and soul; there is no hope of liberation from the world of becoming for man." Bernard Suzanne, "Plato's Phaedrus," http://plato-dialogues.org/tetra_4/phaedrus/plan_sp3.htm#rhetoric (accessed September 6, 2008).

19. All *Phaedo* citations from H. N. Fowler's translation in the Loeb Classical Library.

20. See F. Büchsel, "*Histore* ," in *Theological Dictionary of the New Testament*, ed. G. Kittel and G. Friedrich, trans. G. W. Bromiley (Grand Rapids: Eerdmans, 1965), 3.391–96.

21. Heraclitus reacted against *polymathia* in a way that supports the distinction argued above: "The learning of many things does not reach intelligence; if so, it would have taught Hesiod and Pythagoras, and again Xenophanes and Hecataeus" (frag. 129). True wisdom, according to Heraclitus, consists in understanding the order underlying everything (frag. 41). And this wisdom consists in comprehending the proper relationship that exists between the soul and the world Logos. G. S. Kirk and J. E. Raven, *The Presocratic Philosophers* (London: Oxford University Press, 1966), 204–5.

22. I do not mean that because Plato criticized the physicist program of materialist explanations, he rejected their goal of giving a true account of the causes of coming to be and of perishing. He rejects their explanatory mode (material) and their method or approach to the study of realities. See *Phaedo* 94e, where he discusses his own "makeshift approach" to the problem.

23. Socrates' description of *historia* resembles Guthrie's description of the program and method of philosophy. According to Guthrie, both philosophy and science share the faith that the visible world conceals a rational and intelligible order, that the causes of the natural world are to be sought within its boundaries, and that autonomous human reason is our sole and sufficient instrument for the search. See *A History of Greek Philosophy* (Cambridge: Cambridge University Press, 1962), 1.29.

24. For example, Anaximander explained meteorological phenomena as dependent upon mind: "Winds occur when the finest vapours of the air are separated off and when they are set in motion by congregation; rain occurs from the exhalation that issues upwards from the things beneath the sun, and lightning whenever wind breaks out and clears the clouds" (Kirk and Raven, *Presocratic Philosophers*, 137–38). His

pupil Anaximenes "attributed all the causes of things to infinite air, and did not deny that there were gods, or pass them over in silence; yet he believed not that the air was made by them, but that they arose from air" (*Presocratic Philosophers*, 130).

25. Peter Winch, *The Idea of a Social Science* (London: Routledge & Kegan Paul), 8.

26. John Burnet, *Greek Philosophy*, part 1: *Thales to Plato* (London: Macmillan, 1924), 8–9.

27. This passage (*Phaedo* 99e) does not represent Plato's last word on philosophic technique. Plato did wish to investigate things in themselves. But his method is philosophic because it explicitly states premises and takes a stance of critical analysis in examining theories (*logoi*).

28. "The demand for proof in philosophy seems to have been first occasioned by doctrines that appeared to challenge common sense directly. One such doctrine was the Parmenidean metaphysics of the first half of the fifth century B.C." C. Lejewski, "History of Logic," in *Encyclopedia of Philosophy*, ed. P. Edwards (New York: Macmillan, 1967), 4.513. Kirk and Raven add: "It cannot be too strongly emphasized that before Parmenides and his apparent proof that the senses were completely fallacious . . . gross departures from common sense must only be accepted when the evidence for them is extremely strong" (*Presocratic Philosophers*, 197).

29. Popper, *Conjectures and Refutations*, 89.

30. Plato, *Republic* 8.532c (trans. Benjamin Jowett).

31. On the notion of "progress" in Greek art of the *kouroi*, see Development of Greek Sculpture, http://www.angelfire.com/art/archictecture/articles/dev.htm (accessed September 6, 2008), and Ancient Greek Art, http://witcombe.sbc.edu/ARTHgreece.html (accessed September 6, 2008). For a listing with images of the development of Greek sculptures of youth, see Greek Figure, http://arthist.cla.umn.edu/aict/html/ancient/grfig.html (accessed September 6, 2008).

32. F. B. Tarbell, *History of Greek Art* (New York: Flood & Vincent/London: Macmillan, 1922).

33. Tarbell, *History of Greek Art*, 143.

34. Nigel Spivey, *Understanding Greek Sculpture: Ancient Meanings, Modern Readings* (London: Thames & Hudson, 1996), 24.

35. Spivey, *Understanding Greek Sculpture*, 28.

36. Irish Institute of Hellenic Studies at Athens, http://www.ucc.ie/iihsa/bullthurii.html (accessed September 6, 2008).

37. Walter Pater, *Greek Studies: A Series of Essays* (London: Macmillan, 1910), 239, cited from Gutenberg.org, http://www.gutenberg.org/dirs/etext03/8gsas10.txt (accessed September 6, 2008).

38. Mary Stieber, *The Poetics of Appearance in the Attic Korai* (Austin: University of Texas Press, 2004), 6.

39. Aristotle, *Poetics* 9.1–3, trans. W. Hamilton Fyfe (Cambridge: Harvard University Press, 1927).

40. Otto Michel, "Philosophia," in *Theological Dictionary of the New Testament*, ed. G. Kittel and G. Friedrich, trans. G. W. Bromiley (Grand Rapids: Eerdmans, 1974), 9.172–88.

41. Philip Wheelwright, *The Presocratics* (New York: Odyssey Press, 1966), 26.

42. H. D. F. Kitto, *Form and Meaning in Drama* (London: Methuen, 1956), 243–44.

43. Wheelwright, *Presocratics*, 27.

44. J. E. Sandys, trans., *The Odes of Pindar* (Cambridge: Harvard University Press, 1937), 591n3.

45. All citations from *Sophocles: The Three Theban Plays*, trans. Robert Fagles (New York: Penguin, 1984).

46. David Grene and Richmond Lattimore, eds., *The Complete Greek Tragedies: Sophocles I* (Chicago: University of Chicago Press, 1954), 6.

47. Wallace Stevens, *Harmonium* (New York: Knopf, 1923), 112.

48. C. H. Edwards, *The Historical Development of the Calculus* (New York: Springer, 1979), 10.

49. C. B. Boyer, *The Concepts of the Calculus* New York: Dover, 1959), 21.

50. Boyer, *Concepts of the Calculus*, 23.

51. Boyer, *Concepts of the Calculus*, 45.

52. Boyer, *Concepts of the Calculus*, 53.

53. Empedocles, frag. 109, ca. 440 B.C.E., cited in Aristotle, *Metaphysics* 1000b.

54. Wheelwright, *Presocratics*, 122.

55. See, for example, "The Dynamics of Transference" (1912), in Standard Edition 12.99.

4. SEVEN OF NINE AND FIVE OF NINE

1. *Diagnostic and Statistical Manual of Mental Disorders*, text revision, 4th edition (American Psychiatric Association, 2000), 654.

2. Sophia F. Dziegielewski, *DSM-IV-TR in Action* (New York: Wiley, 2002), 6.

3. Formal Invitations, http://www.formal-invitations.com/invitation-text.html (accessed September 6, 2008).

4. Flannery O'Connor, *Wise Blood* (New York: Farrar, Straus & Giroux, 1952), 105. The first four chapters appeared in magazines in 1948 and 1949.

5. Roy Porter, *Madness: A Brief History* (New York: Oxford University Press, 2003). For a more useful and sober account, see his historical study, *Mind-Forg'd Manacles: A History of Madness in England from the Restoration to the Regency* (London: Athlone, 1987).

6. For example, see Jonathon Andrews and Andrew Scull, *Undertaker of the Mind: John Monro and Mad-Doctoring in Eighteenth-Century England* (Los Angeles: University of California Press, 2001). See also Andrew Wear, ed., *Medicine in Society: Historical Essays* (New York: Cambridge University Press 1992).

7. Beginning with Foucault's *Maladie mentale et personnalité* (1954); *Maladie mentale et psychologie* (1962); *Folie et déraison: Histoire de la folie à l'âge classique*

(1961); *Naissance de la clinique: Une archéologie du régard médical* (1963); *Les Mots et les choses: Une archéologie des sciences humaines* (1966); *Archéologie du savoir* (1969); *Surveiller et punir: Naissance de la prison* (1975); and the three-volume *History of Sexuality* begun in 1976. On Foucault's influence and standing as a historian, see Colin Jones and Roy Porter, eds., *Reassessing Foucault: Power, Medicine, and the Body* (London: Routledge, 1994), 6: "No consensus exists, or indeed is likely, because of the nature of the stakes at issue, to exist for the foreseeable future, between Foucauldians and more classically trained social historians."

8. See H. Tristram Engelhard and Arthur L. Caplan, eds., *Scientific Controversies: Case Studies in the Resolution and Closure of Disputes in Science and Technology* (New York: Cambridge University Press, 1987), chs. 16–18.

9. Irving Bieber, "On Arriving at the American Psychiatric Association Decision on Homosexuality," in *Scientific Controversies*, 417.

10. See Ronald Bayer, *Homosexuality and American Psychiatry: The Politics of Diagnosis* (Princeton: Princeton University Press, 1987). Using a bit of parliamentary chicanery Bieber and others who opposed the outcome of the APA's scientific panels forced the issue to a national referendum among all members of the association: "Ballots were mailed out during April, 1974. Of those responding, only 37% were opposed to the removal of homosexuality from the *DSM-II*. It was a clear endorsement for the change" (144).

11. See Spitzer's extensive research on problems of reliable diagnoses: R. Spitzer, J. Cohen, J. Fleiss, and J. Endicott, "Quantification of Agreement in Psychiatric Diagnosis," *Archives of General Psychiatry* 17 (1967): 83–87; R. Spitzer, J. Endicott, and E. Robins, "Research Diagnostic Criteria: Rationale and Reliability," *Archives of General Psychiatry* 35 (1978): 773–82; R. Spitzer and J. Fleiss, "A Re-analysis of the Reliability of Psychiatric Diagnosis," *British Journal of Psychiatry* 125 (1974): 341–47; R. Spitzer and J. Forman, "DSM-III Field Trials: II. Initial Experience with the Multiaxial System," *American Journal of Psychiatry* 136 (1979): 818–20; R. Spitzer, J. Forman, and J. Nee, "DSM-III Field Trials: I. Initial Interrater Diagnostic Reliability," *American Journal of Psychiatry* 136 (1979): 815–17.

12. Stuart A. Kirk and Herb Kutchins, "The Myth of the Reliability of DSM," *Journal of Mind and Behavior* 15/1–2 (1994): 71–86. See http://home.cc.umanitoba. ca/~mdlee/Teaching/dsmmyth.html (accessed September 6, 2008).

13. Robert Spitzer, "The Diagnostic Status of Homosexuality in DSM-III: A Reformulation of the Issues," in *Scientific Controversies*, 415.

14. "Homosexuality and Psychiatry," http://www.jesus-is-savior.com/Evils%20 in%20America/Sodomy/homosexuality_and_psychiatry.htm (accessed September 6, 2008).

15. Glen O. Gabbard, "Integrated Treatment of Borderline Personality Disorder," in *Psychiatric Times*, http://www.mhsource.com/pt/p960424.html (original April 1996).

16. Alan Morgenstern, "Experiences within a Borderline Syndrome," *International Journal of Psychoanalytic Psychotherapy* 4 (1975): 476–94.

17. Marsha M. Linehan, *Cognitive-Behavioral Treatment of Borderline Personality Disorder* (New York: Guilford, 1993), 16–18.

18. Morgenstern, "Experiences within a Borderline Syndrome," 485–86.

19. Morgenstern, "Experiences within a Borderline Syndrome," 486.

20. See Erving Goffman, *Asylums: Essays on the Social Situation of Mental Patients and Other Inmates* (Garden City, NY: Anchor, 1961); Thomas S. Szasz, *The Myth of Mental Illness* (New York: Hoeber-Harper, 1961); Thomas J. Scheff, *Being Mentally Ill: A Sociological Theory* (Chicago: Aldine, 1966).

21. Thomas Scheff, "Biological Psychiatry and Labeling Theory," in *Being Mentally Ill: A Sociological Theory*, 3rd edition (Hawthorne, NY: Aldine de Gruyter 1991). See Biological Psychiatry and Labeling Theory, http://www.academyanalyticarts.org/scheff.htm (accessed September 6, 2008).

22. See P. Castelnuovo-Tedesco, "Fear of Change as a Source of Resistance in Analysis," *Annual of Psychoanalysis* 14 (1986): 259–72, for a list of common anxieties about beginning an analysis.

23. In *Face of the Deep* (New York; Routledge, 2003), her subtle reflections on *tehom*, Catherine Keller offers a vivid and complex reading of how many ways the hypermasculine God ascribed by orthodoxy to the Hebrew scripture must battle against the feminine and easily dismissed *tehom*.

24. All citations from King James Version: Crosswalk.com, http://bible.crosswalk.com/Lexicons/Hebrew/freqdisp.cgi?book=ps&number=08415&count=11&version=kjv (accessed September 6, 2008).

25. StarTrek.com, http://www.startrek.com/startrek/view/library/character/bio/1112508.html (accessed September 6, 2008).

26. StarTrek.com, http://www.startrek.com/startrek/view/series/VOY/character/1112406.html (accessed September 6, 2008), has biographies of each of the 2,650 main characters.

27. For a list of more than thirty such films, see Humancloning.com, http://www.humancloning.org/movies.htm (accessed September 6, 2008).

28. See Francis Fukuyama, *Our Posthuman Future* (London: Profile, 2002); Ramez Naam, *More Than Human* (New York: Broadway, 2005); Ray Kurzweil and Terry Grossman, *Fantastic Voyage* (New York: Rodal, 2004); James Hughes, *Citizen Cyborg: Why Democratic Societies Must Respond to the Redesigned Human of the Future* (New York: Westview, 2004); and Ray Kurzweil, *The Age of Spiritual Machines: When Computers Exceed Human Intelligence* (New York: Penguin, 1999).

29. Adolph Hitler, *Mein Kampf*, trans. Ralph Manheim (Boston: Houghton Mifflin, 1971) (original 1926).

30. Arthur H. Feiner, "The Dilemma of Integrity," in *Contemporary Psychoanalysis* 11 (1975): 500, quoting Hannah Arendt, *The Human Condition* (Chicago: University of Chicago Press, 1958).

31. Bruno Snell, *The Discovery of the Mind: The Greek Origins of European Thought* (Cambridge: Harvard University Press, 1953).

32. Snell, *Discovery of the Mind*, 123.

33. Snell, *Discovery of the Mind*, 31.

34. Snell, *Discovery of the Mind*, 29.

35. Oedipus Sphinx, http://www.utexas.edu/courses/larrymyth/images/oedipus/4-Oedipus-Sphinx.jpg (accessed September 6, 2008).

36. Unless otherwise noted, all citations are from Sophocles, *The Three Theban Plays: Antigone, Oedipus the King, Oedipus at Colonus*, trans. Robert Fagles (London: Penguin, 1982), 344.

37. *Oedipus the King* 954–63 (Greek text, lines 863–71). On blind rage in Greek combat versus contemporary combat, see Jonathan Shay, *Achilles in Vietnam: Combat Trauma and the Undoing of Character* (New York: Touchstone/Simon & Schuster, 1995). Shay compares the descriptions of "warriors" in the *Iliad* with psychiatric syndromes in modern contexts. See also Hans van Wees's review: "Thus, [Shay] uses the story of Akhilleus as a model instance of the nature and causes of going berserk, regardless of the fact that both Akhilleus' cultural background and his experience of combat differ significantly from that of any American veteran. The 'betrayal of the moral order by a commander,' which he sees as a key to the berserk state, is much less devastating in the case of Akhilleus, who suffers a personal insult in a competitive society where men are expected to compete for respect, violently if necessary, than it is in the modern instances cited, where a commander humiliates or otherwise betrays a soldier whose training has led him to put complete trust in the competence and morality of his superiors" (Hans van Wees, Review of *Achilles in Vietnam* by Jonathan Shay, *Classics Ireland* 5 (1998): 3–21; http://www.ucd.ie/cai/classics-ireland/1998/Wees98.html).

38. Snell, *Discovery of the Mind*, 106.

39. See R. D. Dawe, ed., *Sophocles: Oedipus Rex* (Cambridge: Cambridge University Press, 1982); John Hay, *Oedipus Tyrannus: Lame Knowledge and the Homosporic Womb* (Washington, D.C.: University Press of America, 1978), 27–29. The latter argues that in concrete, mythic systems a name is essential to a person's character, hence "Oedipus" translates as "swollen footed," "know foot," and finally "lame knowledge," the latter designating the castrating, self-blinding action that culminates Oedipus's crusade for insight and wisdom.

40. Snell, *Discovery of the Mind*, 109.

41. All citations from *Thucydides*, trans. Benjamin Jowett, 2nd edition (Oxford: Clarendon, 1900). Pericles' Funeral Oration from *Peloponnesian War* 2.34–46, http://www.wsu.edu:8080/~dee/GREECE/PERICLES.HTM (accessed September 6, 2008).

42. Garry Wills, *Lincoln at Gettysburg: The Words That Remade America* (New York: Simon & Schuster, 1992).

43. Religious studies scholars often defend myths against what they feel are unfair attacks. Yet, in celebrating myths one forfeits a critical edge with which to judge them, except perhaps by aesthetic criteria. In one sense Richard Wagner's transformations of Christian myths into German variants, especially Parsifal, are artistic successes. But, looking backward, we see that when Wagner mythologized the German past it happily coincided with his virulent anti-Semitism, and the latter contributed to the Third Reich's religious heritage. Parsifal was Wagner's Aryan Christ, a better and non-Jewish Christ around whom a newly restored German soul might evolve. The occult qualities in Wagner's drama and Nazi religion are essential to their power, their ability to transcend the burdens of anxiety, of German failures, of divided selves, and to offer unity in the form of an imagined reconciliation.

44. Kernberg makes a similar criticism of Kohution theory about empathy, namely, that it is sometimes reduced to liking the patient and acknowledging only the patient's suffering, like depression and loss. The patient's less attractive feelings, such as rage, envy, and wishes for revenge—which are equally as real—receive less attention. See Otto Kernberg, *Object Relations Theory and Clinical Psychoanalysis* (New York: Aronson 1976).

5 . CANALS ON MARS

1. "Canali" was used by another astronomer, Secchi, years earlier than 1877. See William Sheehan, *Planets and Perception: Telescopic Views and Interpretation, 1609–1909* (Tucson: University of Arizona Press, 1988), 94.

2. Sheehan, *Planets and Perception*, 89.

3. Space art, http://www.bonestell.org/spaceart.html (accessed September 6, 2008).

4. Sheehan, *Planets and Perception*, 182.

5. From David Darling: "Many people felt the haunting, other-worldly allure of the Martian canals and novelists were not slow to weave romantic tales around the theme, further stimulating public interest. Percy Greg, George Griffith, Garrett Serviss, H. G. Wells (most influentially), and others at the end of the nineteenth century, Edgar Rice Burroughs in the early decades of the twentieth, and Ray Bradbury in *The Martian Chronicles* and Robert Heinlein in *Stranger in a Strange Land,* drew inspiration from the Lowellian myth." Internet Encyclopedia of Science, http://www.daviddarling.info/encyclopedia/M/Marscanals.html (accessed September 6, 2008).

6. Sheehan, *Planets and Perception*, 158.

7. See Pyramid Scheme, http://people.ne.mediaone.net/neirr/pyramidscheme.htm (accessed September 6, 2008).

8. Sheehan, *Planets and Perception*, 173, cites A. L. Lowell, *Biography of Percival Lowell* (New York: 1935), 68–69.

9. Eliot Blackwelder, Letters, *Science,* n.s. 29/747 (1909): 659–61 at 659.

10. Classic Encyclopedia, http://9.1911encyclopedia.org/M/MA/MARS.htm (accessed September 6, 2008).

11. Sheehan, *Planets and Perception*, 91.

12. Sheehan, *Planets and Perception*, 187–93.

13. Sheehan, *Planets and Perception*, 193.

14. This replicates the logic that Zeno used in his refutations of the Greek atomists. Sheehan cites another astronomer, Eugène Marie Antoniadi, who responded to Lowell's demand that astronomers decrease the power of their best instruments by shutting down their apertures: "As every reduction in the aperture is accompanied by a corresponding optical distancing of a celestial object, . . . it follows that in order to see a star one must distance it and render it indistinct" (*Planets and Perception*, 249,

citing E. M. Antoniadi, "Les Grand Instruments," *Bulletin de la Société Astronomique de France* 41 [1927]: 146).

15. Percival Lowell, "Mars, I. Atmosphere," *Atlantic Monthly* 75 (May 1895): 594–603; "Mars, II. The Water Problem," *Atlantic Monthly* 75 (June 1895): 749–58; "Mars, III. Canals," *Atlantic Monthly* 76 (July 1895): 106–19; "Mars, IV. Oases," *Atlantic Monthly* 76 (August 1895): 223–35. The quotation is from p. 108 of the July issue. See Mars, http://etext.lib.virginia.edu/etcbin/toccer-new2?id=LowMars. sgm&images=images/modeng&data=/texts/english/modeng/parsed&tag=public&par t=3&division=div1 (accessed September 6, 2008).

16. Sheehan, *Planets and Perception*, 253.

17. Volney P. Gay, *Understanding the Occult: Fragmentation and Repair of the Self* (Minneapolis: Fortress, 1989).

18. Percival Lowell, *Mars and Its Canals* (New York: Macmillan 1906), 365.

19. Karl Popper, *Objective Knowledge: An Evolutionary Approach* (London: Oxford University Press, 1972).

20. William James, *The Varieties of Religious Experience* (New York: Collier, 1902).

21. From Descartes, *Meditations I*, 2.13, cited by Catherine Wilson, "Discourses of Vision in Seventeenth-Century Metaphysics," in *Sites of Vision: The Discursive Construction of Sight in the History of Philosophy*, ed. David Michael Levin (Cambridge: MIT Press, 1999), 117–38 at 124.

22. Irving Langmuir and Robert N. Hall, "Pathological Science," *Physics Today* 42/10 (1989): 36–48. This article is a transcription of a talk Langmuir gave in 1953.

23. For example, Henry H. Bauer, "'Pathological Science' Is Not Scientific Misconduct (Nor Is It Pathological)," in *HYLE — International Journal for Philosophy of Chemistry* 8/1 (2002): 5–20. See HYLE, http://www.hyle.org (accessed September 6, 2008).

24. "Not only did Bohr predict that electrons would occupy specific energy levels, he also predicted that those levels had limits to the number of electrons each could hold. Under Bohr's theory, the maximum capacity of the first (or innermost) electron shell is two electrons. For any element with more than two electrons, the extra electrons will reside in additional electron shells. For example, in the ground state configuration of lithium (which has three electrons) two electrons occupy the first shell and one electron occupies the second shell. To Bohr, the line spectra phenomenon showed that atoms could not emit energy continuously, but only in very precise quantities (he described the energy emitted as *quantized*)." See Atomic Theory, http://www.visionlearning.com/library/module_viewer.php?mid=51 (accessed September 6, 2008).

25. Langmuir and Hall, "Pathological Science," 40.

26. Langmuir and Hall, "Pathological Science," 40.

27. Langmuir and Hall, "Pathological Science," 43.

28. L. V. Beloussov, Faculty of Biology, Moscow State University, Moscow, Russia, and the International Institute of Biophysics, Neuss, Germany. From Abstracts, the 3rd Alexander Gurwitsch Conference on Biophotons and Coherent Systems in Biology, Biophysics and Biotechnology to be held from September 27th to October 2nd,

2004, in Simferopol and Partenit (Crimea, Ukraine). Gurwitsch Conference, http:// gurwitsch.science-center.net/index.php?&i=10&n=pp (accessed September 6, 2008).

29. Mary Jo Nye, "N-rays: An Episode in the History and Psychology of Science," *Historical Studies in the Physical Sciences* 11/1 (1980): 127–56.

30. "On 28 December 1895 Roentgen gave his preliminary report 'Über eine neue Art von Strahlen' to the president of the Wurzburg Physical-Medical Society, accompanied by experimental radiographs and by the image of his wife's hand. By New Year's Day he had sent the printed report to physicist friends across Europe. January saw the world gripped by 'X-ray mania,' and Roentgen acclaimed as the discoverer of a medical miracle. Roentgen, who won the first Nobel prize in physics in 1901, declined to seek patents or proprietary claims on the X-rays, even eschewing eponymous descriptions of his discovery and its applications." James R. Wilson, "Conduct, Misconduct, and Cargo Cult Science," in *Proceedings of the 1997 Winter Simulation Conference: Renaissance Waverly Hotel, Atlanta, Georgia, 7–10 December 1997*, ed. Sigrâun Andradâottir [et al.] (New York: Association for Computing Machinery; Piscataway, NJ: Institute of Electrical and Electronics Engineers, 1997), 1406.

31. Robert W. Wood, "The N-Rays," *Nature* 70 (1904): 530–31. Emotional links, including nationalism and sentiment, seem evident in the French celebration of Blondot and the Russian celebration of Gurwitsch: "Alexander Gavrilovich Gurwitsch (1874–1954), one of the most outstanding biologists of XX century, was in 1918–24 a Professor of Histology in Taurida University. Right there he made his most important discovery of an ultraweak photon emission from the living systems (which gave rise to the biophotonics) and submitted a first sketch of the morphogenetic field theory. The conference is adjusted to the 130 anniversary of Gurwitsch's birthday and, at the same time, to the 100 anniversary of Academician Gleb Frank, a founder of the Institute of Biophysics, Russian Academy of Sciences. Attending in 1920–24 the lectures in Taurida University, Frank was one of the closest Gurwitsch's students and made an important contribution [t]o the biophotonics. By a happy occasion, a house of the first Gurwitsch Laboratory in Simferopol is still preserved and will be decorated by a Memory Desk during the opening of the Conference." Gurwitsch Conference, http:// gurwitsch.science-center.net/index.php?&i=10&n=pp (accessed September 6, 2008).

32. Langmuir and Hall, "Pathological Science," 43.

33. Langmuir and Hall, "Pathological Science," 46.

34. ESP,   http://www.themystica.com/mystica/articles/e/esp_extrasensory_ perception.html (accessed September 6, 2008).

35. J. B. Rhine, *New Frontiers of the Mind: The Story of the Duke Experiments* (New York: Farrar & Rinehart, 1937).

36. Langmuir and Hall, "Pathological Science," 44.

37. See also Wilson, "Conduct, Misconduct, and Cargo Cult Science."

38. We see similar errors—assuming again, this is not mere fraud—in claims about UFOs and such by well-trained persons. See John Mack, *Abduction: Human Encounters with Aliens* (New York: Simon & Schuster 1994).

39. From *Surely You're Joking, Mr. Feynman* (1985), cited in Wilson, "Conduct, Misconduct, and Cargo Cult Science."

40. See David Lowenthal, "The Timeless Past: Some Anglo-American Historical Preconceptions," *Journal of American History* 75/4 (1989): 1263–80; idem, "Letter to the editor," *Perspectives: Newsletter of the American Historical Association* 32 (1994): 17–18; Edward W. Said, *Beginnings: intentIon and Method* (New York: Basic Books, 1976); Brendan Bradshaw, Andrew Hadfield, and Willy Maley, eds., *Representing Ireland: Literature and the Origins of Conflict, 1534–1660* (Cambridge: Cambridge University Press, 1993).

41. Needing to sell a lot of books, Churchill told his publisher how he would compose his *History of the English Speaking Peoples*: "As you know, I wish to give special prominence in the first section of the work to the origin and growth of those institutions, laws and customs and national characteristics which are the common inheritance, or supposed to be, of the English-speaking world. Language and literature play a large part, and indeed these studies should be as it were threaded together by a vivid narrative picking up the dramatic and dominant episodes and by no means undertaking a complete account." Andrew Roberts, in "History Today," http://www.findarticles.com/p/articles/mi_m1373/is_5_52/ai_85677845 (accessed September 6, 2008).

42. All quotations from *Willow*, a story by George Lucas. Screenplay by Bob Dolman. First Draft, Third Revision, November 14, 1986. ©LFL 1986. Lucasfilm LTD. All Rights Reserved. See Willow script, http://www.geocities.com/Area51/Vault/6147/script.txt (accessed September 6, 2008).

43. Macaulay gives an extensive history of the term *Whig* and its counterpart, *Tory*, in ch. 2 of his history: "Under Charles the Second," Names of Whig and Tory, http://www.strecorsoc.org/macaulay/m02e.html (accessed September 6, 2008).

44. All quotations from Thomas Babington Macaulay, *The History of England from the Accession of James II* (London: Macmillan/New York: John Wurtele Lovell, 1882). Published in many editions.

45. Macaulay, *History of England*, 1.13.

46. All Butterfield quotations from Eliohs, http://www.eliohs.unifi.it/testi/900/butterfield (accessed on September 6, 2008).

47. Lowenthal in *Perspectives* (emphasis added). He adds: "Heritage demands faith in a mystique exclusive to devotees. It need not, indeed cannot be proven to outsiders, whom it is meant to mystify or offend. To this end, heritage deploys facts not only unprovable but often demonstrably wrong. Were they not wrong, outsiders could share them. Hence heritage thrives on empirical error. . . . Sharing misinformation excludes those whose own heritage encodes different catechisms. 'Correct' knowledge could not so serve, because it is open to all. What is generally accessible cannot become a criterion of exclusion; only 'false' knowledge can do this. Hence heritage mandates MISreadings of history."

48. For example, John F. Kennedy and his speechwriters echoed the Gettysburg Address and its religious vision of American destiny when they composed Kennedy's famous June 26, 1963, speech in Berlin: "Let me ask you as I close, to lift your eyes beyond the dangers of today, to the hopes of tomorrow, beyond the freedom merely of this city of Berlin, or your country of Germany, to the advance of freedom everywhere, beyond the wall to the day of peace with justice, beyond yourselves and ourselves to

all mankind." The History Place, http://www.historyplace.com/speeches/berliner.htm (accessed September 6, 2008).

6. SEARCHING FOR ESSENCES

1. "We have, however, no right to expect absolute perfection in a part rendered ornamental through sexual selection, any more than we have in a part modified through natural selection for real use; for instance, in that wondrous organ the human eye. And we know what Helmholtz, the highest authority in Europe on the subject, has said about the human eye; that if an optician had sold him an instrument so carelessly made, he would have thought himself fully justified in returning it." Charles Darwin, *The Origin of Species*, ch. 14. The reference to Helmholtz is from *The Field*, May 28, 1870. Hermann von Helmholtz, *Popular Lectures on Scientific Subjects* (London: Longmans, Green, 1873), 219, 227, 269, 390. See also idem, *Handbuch der Physiologischen Optik* (Leipzig: Voss, 1867); idem, *On the Sensations of Tone as a Physiological Theory of Music*, trans. Alexander J. Ellis (repr. New York: Dover, 1954); idem, "The Facts in Perception," in *Hermann von Helmholtz: Philosophical Writings*, ed. Robert S. Cohen and Yehuda Elkana (Boston: Reidel, 1878), 115–63.

2. Alva Noë and Evan Thompson, *Vision and Mind: Selected Readings in the Philosophy of Perception* (Cambridge: MIT Press, 2002), 21.

3. Cornelius Borck, "Visualizing Nerve Cells and Psychical Mechanisms: The Rhetoric of Freud's Illustrations," in *Freud and the Neurosciences: From Brain Research to the Unconscious*, ed. Giselher Guttmann and Inge Scholz-Strasser (Vienna: Verlag der Österreichischen Akademie der Wissenschaften, 1998), 57–86, 60, repr. (Seattle: University of Washington Press, 1999).

4. Borck, "Visualizing Nerve Cells," 63.

5. The original illustration appeared in Sigmund Freud, "Über den Ursprung des Nervus acusticus [On the origin of the auditory nerve]," *Monatsschrift für Ohrenheilkunde sowie für Kehlkopf-, Nasen-, Rachen-Krankheiten*, n.f. 20/8 (1886): 250. Figure 6.1 is from Borck, "Visualizing Nerve Cells," 63.

6. Sigmund Freud, *On Aphasia*, trans. Erwin Stengel (London: Imago, 1953), 72.

7. See Mark Solms and Michael Saling, "On Psychoanalysis and Neuroscience: Freud's Attitude to the Localizationist Tradition," *International Journal of Psychoanalysis* 67 (1986): 397–416. They agree with Borck that *On Aphasia* is the more important text. See Borck, "Visualizing Nerve Cells," 71n15.

8. Sigmund Freud, *The Interpretation of Dreams* (1900), 537, 538, 541 (= Standard Edition vols. 4–5).

9. Freud, *Interpretation of Dreams*, 537.

10. Freud, *Interpretation of Dreams*, 540 (emphasis original).

11. Sigmund Freud, *Das Ich und das Es*, Gesammelte Werke (Frankfurt am Main:

S. Fischer, 1923), 13.252 (= Standard Edition 19.24). The dotted lines illustrate Freud's claim that the ego "rests" upon the id as a germinal disk rests upon the ovum.

12. See Freud, "Notes upon a Mystic-Writing Pad" (1925); *Beyond the Pleasure Principle* (1920), in Standard Edition 18.7; *Group Psychology and the Analysis of the Ego* (1921), in Standard Edition 18.69; and *The Ego and the Id* (1923), in Standard Edition 19.3.

13. Borck, "Visualizing Nerve Cells," 85.

14. See Sigmund Freud, "A Project for a Scientific Psychology," in *The Origins of Psycho-Analysis*, trans. Eric Mosbacher and James Strachey (New York: Basic Books, 1895), 347–445. Revised and reprinted in Standard Edition 1.175, 365–67. For other comments on these passages see Akira Mizuta Lippit, "Virtual Annihilation: Optics, VR, and the Discourse of Subjectivity," *Criticism* 36/4 (1994): 595.

15. Even then, problems emerged: "By the early nineteenth century, the camera lost its efficacy because the subject had, in essence, become the camera: it had metamorphosed into a representation of the simulated world. During the nineteenth century a series of optical devices—John Paris's 'Thaumatrope,' David Brewster's 'Kaleidoscope,' Joseph Antoine Ferdinand Plateau's 'Phenakistiscope,' Eadweard [*sic*] Muybridge's 'Zoopraxiscope,' to name only a few—gradually moved the screen of representation from the world to the psyche." Lippit, "Virtual Annihilation."

16. All Plato citations from *Plato*, trans. Paul Shorey (Cambridge: Harvard University Press, 1969), vols. 5–6.

17. Novalis (Friedrich von Hardenberg), *Blütenstaub* (original 1797), trans. Charles Rosen in *The Classical Style: Haydn, Mozart, Beethoven* (New York: Norton, 1971). Among other aphorisms by Novalis: "The person is a sun, his senses are its planets"; "the person is himself the largest secret, even—the solution of this endless task in the world's story"; "we are near awakening when we dream that we dream."

18. As I argue in *Joy and the Objects of Psychoanalysis* (Albany: State University of New York Press, 2001).

19. All Aristotle citations from *Aristotle*, trans. Hugh Tredennick (Cambridge: Harvard University Press/London: Heinemann), vols. 17–18. Concerning Aristotle and the idealization of vision, see Hans Jonas, *The Phenomenon of Life: Toward a Philosophical Biology* (New York: Harper & Row, 1966). On "sight" and philosophy, see David Michael Levin, ed., *Sites of Vision: The Discursive Construction of Sight in the History of Philosophy* (Cambridge: MIT Press, 1997).

20. Claude Lévi-Strauss, *Tristes Tropiques*, trans. John Russell (New York: Criterion, 1961), 6.

21. Roger T. Ames, *Sun-Tzu: The Art of Warfare* (New York: Ballantine, 1993), 49.

22. For example, on Archimedes, see Archimedes and the Computation of Pi, http://www.math.utah.edu/~alfeld/Archimedes/Archimedes.html (accessed September 6, 2008). On Greek mathematics see also Wilbur R. Knorr, *The Evolution of the Euclidean Elements: A Study of the Theory of Incommensurable Magnitudes and Its Significance for Early Greek Geometry* (Boston: Reidel, 1975).

23. Sigmund Freud, "An Outline of Psycho-Analysis" (1940), in Standard Edition 23.196.

24. Jacob A. Arlow and Charles Brenner, "The Future of Psychoanalysis," *Psychoanalytic Quarterly* 57 (1988): 1–14 at 9. See also Charles Brenner, "Psychoanalysis and Science," *Journal of the American Psychoanalytic Association* 16 (1968): 675–96.

25. Michael Franz Basch, "Theory Formation in Chapter VII: A Critique," *Journal of the American Psychoanalytic Association* 24 (1976): 61–100; Donald P. Spence, "The Rhetorical Voice of Psychoanalysis," *Journal of the American Psychoanalytic Association* 38 (1990): 579–603.

26. Brenner, "Psychoanalysis and Science," 686.

27. Seeming to prove Spence's point are Whitehead's speculations about Freud's choice of the metaphor "telescope" in his discussion of dreams: "One is immediately struck by the metaphorical selection of a telescope. Telescopes are frequently long, thin, round objects which may have remarkable expansive potentialities. Clearly a phallic imagery suggests itself in this psychological metaphor. Thus Freud's reaction to the awe inspiring feminine dream is the 'erection of a psychologic scaffolding' (p. 568) of phallic conformation." Clay C. Whitehead, "Additional Aspects of the Freudian-Kleinian Controversy: Towards a 'Psychoanalysis' of Psychoanalysis," *International Journal of Psychoanalysis* 56 (1975): 383–96 at 389.

28. See also E. Nagel, "Methodological Issues in Psychoanalytic Theory," in *Psychoanalysis, Scientific Method, and Philosophy*, ed. S. Hook (New York: New York University Press, 1959), 38–56.

29. Austrian Ludwig Wittgenstein Society, Kirchberg am Wechsel, Austria. See image at http://www.sbg.ac.at/phs/alws/pics/wittgenstein2-big.jpg.

30. Ludwig Wittgenstein, *Tractatus Logico-Philosophicus*, German text with English translation by C. K. Ogden and introduction by Bertrand Russell (London: Routledge and Kegan Paul, 1922). Revised translation by Jonathan Laventhol at http://www.kfs.org/~jonathan/witt/t411en.html (accessed September 6, 2008).

31. Laventhol translation, http://www.kfs.org/~jonathan/witt/t446en.html.

32. Ogden translation, 69.

33. Laventhol translation, http://www.kfs.org/~jonathan/witt/ten.html.

34. Note, too, Wittgenstein's resonance with advances in artificial languages and computation. See Charles Petzold, *Code: The Hidden Language of Computer Hardware and Software* (Seattle: Microsoft, 2000).

35. Ludwig Wittgenstein, *Philosophical Investigations*, trans. G. E. M. Anscombe (New York: Macmillan, 1953).

36. Wittgenstein, *Philosophical Investigations*, 18e, 42e (emphasis original).

37. Wittgenstein, *Tractatus Logico-Philosophicus*, no. 5.5563.

38. Wittgenstein, *Philosophical Investigations*, 44e.

39. Wittgenstein, *Philosophical Investigations*, 46e.

40. For additional comments on Wittgenstein's relevance to psychoanalysis and the question of the self, see Peter Fonagy et al., "Measuring the Ghost in the Nursery: An Empirical Study of the Relation Between Parents' Mental Representations of Childhood Experiences and Their Infants' Security of Attachment," *Journal of the American Psychoanalytic Association* 41 (1993): 957–89; Marcia Cavell, "Solipsism and Community: Two Concepts of Mind in Psychoanalysis," *Psychoanalysis and*

Contemporary Thought 11 (1988): 587–613; Donald Davidson, *Inquiries into Truth and Interpretation* (Oxford: Oxford University Press, 1984), esp. ch. 17: "What Metaphors Mean"; and idem, "A Coherence Theory of Truth and Knowledge," in *Truth and Interpretation: Perspectives on the Philosophy of Donald Davidson*, ed. Ernest LePore (Oxford: Blackwell, 1986), 307–19.

41. Wittgenstein, *Philosophical Investigations*, 48e.

42. Freud, "The Question of a Weltanschauung" (1933), in Standard Edition 22.160.

43. Freud, "Dostoevsky and Parricide" (1928), in Standard Edition 21.175.

44. Fyodor Dostoevsky, *The Brothers Karamazov*, trans. Constance Garnett (New York: Modern Library, 1943), part IV, book XII, ch. 9.

45. Dostoevsky, *Brothers Karamazov*, part IV, book XII, ch. 10.

46. "The Thrill Is Gone" lyrics, http://www.bluesforpeace.com/lyrics/thrill-is-gone.htm (accessed September 6, 2008).

47. According to Eric Predoehl: "The original 'Louie Louie' was written in 1955 by Richard Berry and released as a single in 1957 on Flip Records. Recorded with the Pharaohs, Richard created a catchy, somewhat calypso ditty that was originally intended as the B-side for his recording of 'You Are My Sunshine.'" Louie-Louie, http://www.louielouie.net/03-richardberry.htm (accessed September 6, 2008).

48. Pat Righelato, "Wallace Stevens," in *American Poetry: The Modernist Ideal*, ed. Clive Bloom and Brian Docherty (New York: St. Martin's, 1995). Excerpted at http://www.translatum.gr/forum/index.php?topic=6178.0.

49. Glen MacLeod, *Wallace Stevens and Modern Art: From the Armory Show to Abstract Expressionism* (New Haven: Yale University Press, 1993), 20–22. Excerpted at Anecdote of the Jar, http://www.english.uiuc.edu/maps/poets/s_z/stevens/jar.htm (accessed September 6, 2008).

50. Frank Lentricchia, *Ariel and the Police: Michel Foucault, William James, and Wallace Stevens* (Madison: University of Wisconsin Press, 1988).

51. B. J. Leggett, *Early Stevens: The Nietzschean Intertext* (Durham: Duke University Press, 1992), 205.

52. See Swammerdam, http://www.janswammerdam.net/nerve.html (accessed September 6, 2008).

53. See Edward G. Ruestow, *The Microscope in the Dutch Republic: The Shaping of Discovery* (New York: Cambridge University Press, 1996).

54. Percival Lowell, *Mars* (Boston: Houghton, Mifflin, 1895); *Mars and Its Canals* (New York: Macmillan, 1906); and *Mars as the Abode of Life* (New York: Macmillan, 1908).

55. Freud, *Interpretation of Dreams*, 111n1. See also: "Even in the best interpreted dreams, there is often a place [*eine Stelle*] that must be left in the dark, because in the process of interpreting one notices a tangle of dream-thoughts arising [*anhebt*] which resists unraveling but has also made no further contributions [*keine weiteren Beiträge*] to the dream-content. This, then, is the navel of the dream, the place where it straddles the unknown [*dem Unerkannten aufsitzt*]. The dream-thoughts, to which interpretation leads one, are necessarily interminable [*ohne Abschluss*] and branch

out on all sides into the netlike entanglement [*in die netzartige Verstrickung*] of our world of thought. Out of one of the denser places in this meshwork, the dream-wish rises [*erhebt sich*] like a mushroom out of its mycelium" (525 = trans. Weber [1982], 75). Stanley J. Coen, "The Passions and Perils of Interpretation (of Dreams and Texts): An Appreciation of Erik Erikson's Dream Specimen Paper," *International Journal of Psychoanalysis* 77 (1996): 537–48.

56. "Colon [:] in Greek means member, parts phallus (penis). Indeed when we use a colon in punctuation, it divides a sentence into parts or members." Principles of Investigation, http://www.apol.net/dightonrock/basic_principals_of_investigatin.htm (accessed September 6, 2008).

57. Paul Bach-y-Rita and Stephan W. Kercel, "Sensory Substitution and the Human-Machine Interface," *Trends in Cognitive Sciences* 7/12 (2003): 541–46.

58. See Paul Bach-y-Rita, "Theoretical Basis for Brain Plasticity after a TBI," *Brain-Injury* 17/8 (2003): 643–51; and G. L. Aiello, "Multidimensional Electrocutaneous Stimulation," *IEEE Transactions of Rehabilitation Engineering* 6 (1998): 1–7.

59. See Paul Bach-y-Rita, Mitchell E. Tyler, and Kurt-A Kaczmarek, "Seeing with the Brain," *International Journal of Human-Computer Interaction* 15/2 (2003): 285–95.

60. Bach-y-Rita and Kercel cite M. J. Morgan, *Molyneux's Question* (London: Cambridge University Press, 1977).

61. Bach-y-Rita and Kercel cite Paul Bach-y-Rita, *Nonsynaptic Diffusion Neurotransmission and Late Brain Reorganization* (New York: Demos-Vermande, 1995).

62. Bach-y-Rita and Kercel cite P. B. L. Meijer, "An Experimental System for Auditory Image Representations," *IEEE Transactions of Rehabilitation Engineering* 39 (1992): 112–21.

63. Bach-y-Rita and Kercel, "Sensory Substitution and the Human-Machine Interface," 544.

64. J. J. Gibson, *The Ecological Approach to Visual Perception* (Boston: Houghton-Mifflin, 1979). For a recent philosophical attempt that pursues Gibson's project, see Harwood Fisher, "Categories and Embodied Objects: The Subjective Self and the Psychologist Within Natural Psychology," *Theory and Psychology* 13/2 (2003): 239–62.

65. For a conjecture as to how perception and environment interact to make consciousness possible, see J. Kevin O'Regan and Alva Noë, "What It Is like to See: A Sensorimotor Theory of Perceptual Experience," *Synthese* 129/1 (2001): 79–103. They propose a way to bridge "the gap between physical processes in the brain and the 'felt' aspect of sensory experience. The approach is based on the idea that experience is not generated by brain processes themselves, but rather is constituted by the way these brain processes enable a particular form of 'give-and-take' between the perceiver and the environment" (79).

66. For additional support of this interactive view of perception, see F. J. Varela, E. Thompson, and E. Rosch, *The Embodied Mind: Cognitive Science and Human Experience* (Cambridge: MIT Press, 1991). For an excellent overview of these debates, see Noë and Thompson, *Vision and Mind*. See also Hans Jonas, *The Phenomenon of Life: Toward a Philosophical Biology* (New York: Harper & Row, 1966).

67. See F. A. Hayek, *The Road to Serfdom* (Chicago: University of Chicago Press, 1944); idem, *Individualism and Economic Order* (Chicago: University of Chicago Press, 1948); idem, *That Fatal Conceit: The Errors of Socialism* (Chicago: University of Chicago Press, 1988).

68. Descartes, http://www.wsu.edu:8080/~wldciv/world_civ_reader/world_civ_reader_2/descartes.html (accessed September 6, 2008); see Hayek, *Individualism and Economic Order*, 9–11.

69. Hayek, *Individualism and Economic Order*, 8.

70. Hayek, *Individualism and Economic Order*, 8.

71. Hayek, *Individualism and Economic Order*, 15.

72. Hayek, *Individualism and Economic Order*, 19.

7. HIGH ART AND THE POWER TO GUESS THE UNSEEN FROM THE SEEN

1. All Ruskin citations from *The Works of John Ruskin*, ed. E. T. Cook and A. Wedderburn, 39 vols. (London: George Allen, 1903–12).

2. George P. Landow, *The Aesthetic and Critical Theories of John Ruskin* (Princeton: Princeton University Press, 1971), online at http://www.victorianweb.org/authors/ruskin/atheories/contents.html.

3. *Works of John Ruskin*, 6.206.

4. By folk etymology, the word *demijohn* is derived from French *dame-jeanne*, literally, Lady Jane (1806), a large narrow-necked bottle usually enclosed in wickerwork. A *tierce* is a measure of liquid capacity, equal to a third of a pipe, or 42 gallons (159 liters). We find "11 tierce" mentioned in "An Inventory of Articles at Mount Vernon, 1810" in the estate of George Washington; http://gwpapers.virginia.edu/documents/inventory/list.html (accessed September 6, 2008).

5. Sigmund Freud, "Mourning and Melancholia" (1917), in Standard Edition 14.239–60.

6. Freud, "Mourning and Melancholia," 245.

7. Freud, "Mourning and Melancholia," 245.

8. Melanie Klien, "Mourning and Its Relation to Manic-Depressive States," *International Journal of Psychoanalysis* 21 (1940): 125–53 at 136.

9. Klien, "Mourning and Its Relation to Manic-Depressive States," 136.

10. Thus, Klein adds in a footnote: "These facts I think go some way towards answering Freud's question which I have quoted at the beginning of this paper: 'Why this process of carrying out the behest of reality bit by bit, which is in the nature of a compromise, should be so extraordinarily painful is not at all easy to explain in terms of mental economics. It is worth noting that this pain seems natural to us'" ("Mourning and Its Relation to Manic-Depressive States," 136n15).

11. Freud, "Hysteria" (1888), in Standard Edition 1.39–59 at 57.

12. Freud, "Hysteria," 57.

13. Freud, "Preface to the Translation of Bernheim's *Suggestion*" (1888), in Standard Edition 1.72–87 at 84.

14. Freud, "Some Points for a Comparative Study of Organic and Hysterical Motor Paralyses" (1893), in Standard Edition 1.155–72 at 169.

15. Freud, "Organic and Hysterical Paralyses," 172.

16. Robert Gaines, "Detachment and Continuity," *Contemporary Psychoanalysis* 33 (1997): 549–71 at 551.

17. Salman Akhtar, "Mental Pain and the Cultural Ointment of Poetry," *International Journal of Psychoanalysis* 81 (2000): 229–43 at 236.

18. Akhtar, "Mental Pain," 236n11: "Poets frequently find out what they were writing about only after they have completed a poem. The Nobel prize–winning Irish poet Seamus Heaney's comment that in writing poetry, there is a 'movement from delight to wisdom and not vice versa' (1995, 5) refers to this very point." The Seamus Heaney citation is from *The Redress of Poetry* (New York: Farrar, Straus & Giroux, 1995).

19. On the sources of Freud's formulations, see Martine Lussier Paris, "Mourning and Melancholia," *International Journal of Psychoanalysis* 81 (2000): 667–86.

20. Gaines, "Detachment and Continuity," 551.

21. At the end of the first season, Tony, the mobster-patient, laments that "cunnilingus and psychiatry brought us to this!" While very funny, David Chase, creator of *The Sopranos*, admits that the line is less than artistic since it ascribes falsely to Tony an insight that he does not have.

22. See Peter Winch, *The Idea of a Social Science* (London: Routledge & Kegan Paul, 1958). Empathy is the ability to comprehend the lived world of the other, how she or he perceives a world of affordances.

23. Henry James, "The Art of Fiction," *Longman's Magazine* 4 (September 1884); repr. in James's *Partial Portraits* (London: Macmillan, 1888), 65–66 (emphasis added). The archaic word *ell* derives from the Latin word *ulna*. It is an anthropomorphic measure of about 45 inches: "The word ell seems to have been variously taken to represent the distance from the elbow or from the shoulder to the wrist or to the finger-tips, while in some cases a 'double ell' has superseded the original measure, and has taken its name. English ell = 45 in. Scots = 37.2 in. Flemish = 27 in." See Measurement in the Middle Ages, http://www.personal.utulsa.edu/~marc-carlson/history/measure.html (accessed September 6, 2008).

24. Volney P. Gay, *Freud on Sublimation* (Albany: State University of New York Press, 1992).

25. Pinchas Noy, "Symbolism and Mental Representation," *Annual of Psychoanalysis* 1 (1973): 125–58.

26. J. L. Austin, *Sense and Sensibilia* (Oxford: Clarendon, 1962), 2.

27. J. J. Gibson, *The Ecological Approach to Visual Perception* (Boston: Houghton-Mifflin, 1979).

28. Sigmund Freud, "A Project for a Scientific Psychology," in *The Origins of Psycho-Analysis*, trans. Eric Mosbacher and James Strachey (New York: Basic Books, 1895), 367ff. = Standard Edition 1.283–397.

29. *Blow Up*, http://ekatocato.hippy.jp/gazou/bu8.jpg (accessed September 6, 2008).

30. Thomas Beltzer, "La Mano Negra: Julio Cortázar and His Influence on Cinema," http://www.sensesofcinema.com/contents/05/35/cortazar.htm (accessed September 6, 2008).

31. Julio Cortázar, cited in http://www.kirjasto.sci.fi/cortaz.htm (accessed September 6, 2008).

32. *The Poetical Works of John Keats* (Boston: Little, Brown, 1863), 216.

33. Interview with Julio Cortázar, by Evelyn Picon Garfield, http://www.center forbookculture.org/interviews/interview_cortazar.html (accessed September 6, 2008). This interview was conducted in Saignon, France, July 10–13, 1973, and is excerpted from Garfield's *Cortázar por Cortázar* (Jalapa, Mexico: Universidad Veracruzana, 1978). From *Review of Contemporary Fiction* 3.3 (Fall 1983).

34. Michelangelo Antonioni, *Screenplays* (New York: Orion, 1963), ix.

35. Antonioni, *Screenplays*, xiii.

36. Antonioni, *Screenplays*, xiv.

37. Nurse Diesel, www.heavenandearthandyou.com/props.htm (accessed September 6, 2008).

38. Thus, a 35 mm print (= 35 mm x 35 mm = 1,225 sq. mm) must be enlarged by a factor of 816.3 to become 10 meters square (= 1,000 mm x 1,000 mm = 1,000,000 sq. mm). On specifications for 35 mm film, see http://photography.cicada.com/zs/tables/basic.html. On specifications for newsprint photos, see http://www.epixel.com/DK/dk/basics/pixel.htm (accessed September 6, 2008), http://www.designer-info.com/master.htm?http://www.designer-info.com/DTP/understanding_halftones.htm (accessed September 6, 2008).

39. The Indo-European stem *men* has a role in words that pertain to thinking or states of mind such as *mind, mentor, mentation, mandarin*, and *mantra*.

40. "Particularly curious are those intended for the relief of pregnant women and that of children. For instance, a well-known practise among them is 'Bleigiessen,' or what may be termed 'plumbomancy,' which is divination from the forms assumed by molten lead dropped into water. This is resorted to in cases in which illness of pregnant women or that of children is due to fright, to find out what object was the cause of the alarm. A medicine-woman, muttering a psalm or an incantation, throws molten lead into a vessel full of water, and from the resemblance of the form thus assumed by the metal to a particular animal, she divines that the cause of fright was a cat, a dog, a horse, etc." Herman Rosenthal and Alexander Harkavy, "Babski Refues," cited from http://www.jewishencyclopedia.com/view.jsp?artid=50&letter=B (accessed September 6, 2008).

41. Francis Ford Coppola says that he recorded the scene in Union Square in San Francisco using the kinds of devices seen in the movie. All references to Coppola's discussion are from his comments on the DVD version of the film, release date: December 12, 2000, Paramount.

42. E. Cobham Brewer's 1898 *Dictionary of Phrase and Fable* states: "The membrane on the heads of some new-born infants, supposed to be a charm against death by drowning. To be born with a caul was with the Romans tantamount to our phrase,

'To be born with a silver spoon in one's mouth,' meaning 'born to good luck.'" Cited from http://www.bartleby.com/81/3246.html (accessed September 6, 2008).

43. Charles Dickens, *David Copperfield*, ch. 1: "I Am Born": "I was born with a caul, which was advertised for sale, in the newspapers, at the low price of fifteen guineas. Whether sea-going people were short of money about that time, or were short of faith and preferred cork jackets, I don't know; all I know is, that there was but one solitary bidding, and that was from an attorney connected with the bill-broking business, who offered two pounds in cash, and the balance in sherry, but declined to be guaranteed from drowning on any higher bargain." Cited from ftp://ibiblio.org/pub/docs/books/gutenberg/etext96/cprfd10.txt (accessed September 6, 2008).

44. For Harry's conversation, see http://ruthlessreviews.com/pics/conversation1.gif (accessed September 6, 2008).

45. All quotations from *The Conversation* are from Tim Dirks, "The Conversation," http://www.filmsite.org/conv.html (accessed September 6, 2008).

46. All of the quotations in this section are from Gary Taubes, "Physics: The One That Got Away?" *Science* 275 (1997): 148–51 (in *News & Comment* [unpaginated]).

47. A KeV equals one thousand electron volts. An electron volt (abbreviated eV) is "a unit of energy used to describe the total energy carried by a particle or photon. The energy acquired by an electron when it accelerates through a potential difference of 1 volt in a vacuum. 1 eV = 1.6 x 10-12." See Solar Physics Glossary, http://hesperia.gsfc.nasa.gov/sftheory/glossary.htm#EV (accessed September 6, 2008).

48. For an example of Rhine's methods, see J. B. Rhine, *New Frontiers of the Mind: The Story of the Duke Experiments* (New York: Farrar & Rinehart, 1937).

49. Cited in the Crossroads Initiative, http://www.crossroadsinitiative.com/resource_info/198.html (accessed September 6, 2008).

50. *Catechism of the Catholic Church*, 2nd ed., cited from http://www.scborromeo.org/ccc/p2s2c1a1.htm#1238 (accessed September 6, 2008).

51. Jir Takei and Marc P. Kean, *Sakuteiki: Visions of the Japanese Garden* (Boston: Tuttle, 2001).

52. John Donne, "Expostulation" #19 in *Devotions* (Ann Arbor: University of Michigan, 1959), 124. Thanks to Professor Fred Shriver, General Theological Seminary, New York City, for showing me this text.

53. Donne, "Expostulation," 125.

54. Donne, "Expostulation," 125.

55. Donne, "Expostulation," 124.

56. André Green, "The Illusions of Common Ground and Mythical Pluralism," trans. Sophie Leighton, *International Journal of Psychoanalysis* 86/3 (2005): 627–32 at 631.

57. The *PEP Archive CD-ROM* (Psychoanalytic Electronic Publishing, 2003) includes the full-text of eleven psychoanalytic journals from 1920 to 2000 and the full text of more than thirty-five thousand articles and books.

58. Ernest Jones, *The Life and Work of Sigmund Freud*, vol. 2: *Years of Maturity, 1901–1919* (New York: Basic Books, 1955), 423.

59. Green, "Illusions of Common Ground," 632 (emphasis original).

60. Green, "Illusions of Common Ground," 628 (emphasis original).

61. Green, "Illusions of Common Ground," 629.

62. We hear excerpts from the Concerto for Flute and Harp, K. 299; 2nd movement Symphony No. 29, K. 201; 1st movement Concerto for two pianos, K. 365; 3rd movement Symphonie Concertante, K. 364; 1st movement Mass in C Minor, K. 427; and "Kyrie." See http://www.amadeusimmortal.com/movie/musicguide.php (accessed September 6, 2008). See the movie script, 54–55, at http://www.amadeus immortal.com/movie/Amadeus.pdf (accessed September 6, 2008) and http://sfy.ru/sfy.html?script=amadeus (accessed September 6, 2008).

63. See Volkmar Braunbehrens, *Maligned Master: The Real Story of Antonio Salieri* (New York: Fromm, 1992).

64. For example, see Ellen Schrecker, ed., *Cold War Triumphalism: The Misuse of History After the Fall of Communism* (New York: New Press, 2004).

INDEX